WHAT PEOPLE ARE SAYING ABOUT

SAFE PLANET

By focusing on energy issues, John Cowsill shows how the global political and economic structures that make up the capitalist system are responsible for the multiple crises currently affecting both people and planet and identifies the agency which has the power to get us out of this mess.
Ellen Clifford, Disabled People Against Cuts

This is a timely and thought-provoking book. Coming from an unashamedly anti-capitalist position, it argues that a sustained programme of investment in renewables is the way to deal with our energy crisis. Not everybody will agree with all the arguments put forward in the book. But everybody will be challenged by it.
John Stewart, Chair, Campaign Against Climate Change (CCC) (Personal Capacity) and Chair, Heathrow Association for the Control of Aircraft Noise (HACAN) (Personal Capacity)

I have read the book and was impressed ... I would say that this book would appeal to the interested generalist.
Mike Thompson, MSc Course Director at Anglia Ruskin University

Safe Planet

Renewable Energy plus Workers' Power

Safe Planet

Renewable Energy plus Workers' Power

John Cowsill

Winchester, UK
Washington, USA

First published by Earth Books, 2014
Earth Books is an imprint of John Hunt Publishing Ltd., Laurel House, Station Approach, Alresford, Hants, SO24 9JH, UK
office1@jhpbooks.net
www.johnhuntpublishing.com
www.earth-books.net

For distributor details and how to order please visit the 'Ordering' section on our website.

ISBN: 978 1 78099 682 0

A CIP catalogue record for this book is available from the British Library.

Design: Stuart Davies

Printed and bound by CPI Group (UK) Ltd, Croydon, CR0 4YY

We operate a distinctive and ethical publishing philosophy in all areas of our business, from our global network of authors to production and worldwide distribution.

CONTENTS

Acknowledgments

The following friends and comrades read the book and gave me very valuable feedback:

Ellen Clifford, Martin Empson, Suzanne Jeffery, John Stewart, Mike Thompson.

Thanks to you all for giving your time and support.

Thanks also to all at John Hunt Publishing for publishing the book, and for all your hard work.

Safe Planet

Introduction

The capitalist system is in a seemingly never ending p
economic and ecological crisis. In the last few years, di
have been toppled, large chunks of the banking sector have
and are going bankrupt and the earth's ecosystem is grad
being degraded. The planet's life support system is b
destroyed by the very same mechanisms that are to blame for
financial and economic instability[1].

Billions of ordinary people are humiliated by unemploymen
debt and insecurity while those lucky enough to hav
employment have had their wages cut. Coincidentally, the
fortunes of a tiny minority of the world's population, the rich, the
so called 'high net worth individuals' have risen to new heights.
This wealth transfer is happening while the ecosystem degradation is blighting and destroying the lives of some of the
poorest and most vulnerable people as the extreme weather
floods and droughts brought on by global temperature ris
wreak their havoc. I aim to show, in this book, how the tw
aspects of the crisis have the same cause.

One of the less obvious effects of the crisis is the way it in
ences ideas and opinions. So, for example, I am interested ir
potential for renewable energy. All the time I am confronted
ideas like: 'Humanity cannot provide the energy and po
needs using renewable sources of energy alone'; 'Rene
power is uneconomic'; 'It is impossible to provide the po
need without nuclear'; 'Windpower will always need
from conventional generation'.

I am writing this book, firstly, as a challenge to thes
notions and assertions about renewable energy. I
showing, in some detail, how two example states, Ca

the UK, can provide all the power and energy they need using renewable sources alone. I argue that if it is possible in those high consumption states it is achievable everywhere. Meeting this challenge is the first task of the book and I have attempted to do this – in Part 1 – starting with two sets of real meteorological data from the two geographical locations.

However, completing that task isn't enough because that technical explanation does not explain the lack of progress towards producing an energy infrastructure based on renewable sources of energy. So a second task is to attempt to understand and explain what is behind the lack of progress towards a more rational energy infrastructure.

There are many aspects to the crisis. Food prices are sky-rocketing, there is a homelessness crisis, water supplies are under threat in many parts of the world and there is a major loss of bio-diversity. Those are just a few of the many problems. The list of problems seems endless. Of course, I am not able to deal in any detail at all with all of these things. However, I believe, and attempt to show here, that all the crises are linked, and in the second part of the book I try to address some of these issues, and show that the solution to the climate crisis, and to the financial crisis, is also the solution to all the other crises such as those in agriculture and construction etc.

The book is not being written from an unbiased position. It is not possible to be unbiased in this very unbalanced world. Those who say they comment on world politics and economics from a 'balanced', neutral position, are either being extremely naïve or are disingenuous.

I am on the side of the vulnerable and oppressed. I am on the side of all those who are fighting back. This book is written from the point of view of the 99% and against the 1%. It is against the bankers and their friends with their million dollar bonuses and against the faceless bosses who make millions from selling arms and promoting war - the death trade. It is on the side of disabled

For Mum and Dad, Jean, Chris, Kate and Jake.

people fighting cuts, it is on the side of the Occupy movement, it is on the side of the Indignados (the outraged). It is for those fighting against sexual harassment. It is for those fighting sexism, racism and homophobia and all forms of discrimination. It is for workers and unemployed workers everywhere standing up and fighting for a better world.

Having said which side I am on, I think it is necessary to be objective, to be scientific, to face the facts, in short, to tell the truth about what is happening as I see it. In the chapters on how it is possible to provide all our energy from renewables, this is fairly simple. I take some wind speed data, and some data about the strength of the sun at specific times, and match it up with the power demand at those times. By doing that I calculate how many wind turbines, how much solar panelling and how much storage will be required to satisfy the demand.

In Part 2 and the chapter on 'the enemy within' it is a little more difficult. It is impossible to get hold of all the relevant information on the fossil fuel and nuclear power companies, the banks etc, as these companies value their 'commercial privacy' extremely highly and keep a lot of important information away from the public. However, there is more than enough available in the public domain to achieve what I want to achieve.

I sift through some of this evidence to show firstly how the fossil fuel companies are a major part of the problem – a major ingredient in the twin crises – but I also try to show how the men and women currently working within these organisations can be a major part of the solutions. I do this by examining the record of these corporations to see to what extent the oil, coal and nuclear corporations – and their closely linked governments – are central to the running of the global economy, but to see to what extent also, they are dishing the dirt on us and on the planet.

Five years ago, at the start of the current financial crisis, massive fraudulent activity was discovered at the very heart of the system. When Lehman Brothers, one of the world's largest banking groups and AIG, which was at the time, the world's largest insurance company, were being consigned to the dustbin of history, worried and panic-stricken politicians, like the British Prime Minister, Gordon Brown, and his chancellor, Alistair Darling, could be seen on television looking tired and dishevelled due to having been up all night 'saving the world' as Brown described it, talking bluntly about the failure of capitalism itself.

Karl Marx, the author of 'Capital', a bit of an expert about capitalism, had he still been alive, would have been laughing out loud at the thought of that.

In the months and years following the collapse, governments around the world have printed and spent billions of Dollars, Euros and Pounds in an attempt to bail out the system. They are trying to do this without upsetting the senseless luxury, extravagance and ridiculous wealth of those who are the prime beneficiaries of this unbalanced, unequal world. And they are doing it by squeezing harder and harder those who are already squeezed dry, the 99%, the vast majority, the people whose work creates all the wealth.

Part 2 then, is firstly about what is all wrong with things and how it got that way.

But it is also about solutions – it identifies the agency for change.

All across the world, in Egypt, Tunisia, Syria, in the United States, in China, Turkey, Spain, Brazil and in many other countries, people are rising up to demand change. The capitalist system is showing signs of cracking under the strain.

All round the world students are marching, workers are striking and people everywhere are challenging the old order. The 'Arab Spring' uprisings began, in Tunisia, as a response to the austerity measures which were imposed on the population by a brutal dictator. That dictator has now gone as has the Egyptian people's tormentor, Mubarak. But the story is still being written. The confusion and horror of the war in Syria which began as a reaction by the existing regime when part of the population also rose up to remove their oppressors must be levelled at the capitalist system. Those weeping, lethal chemical weapons unleashed upon now dead innocents were developed in Britain (in Cornwall, a quiet and peaceful backwater of England!) and Germany and sold to Syria by faceless arms dealers. It also serves as a timely reminder to would-be revolutionaries; those who half make a revolution are destined to fail.

Some capitalists and apologists for the system – most likely because they fear the wrath of those who are suffering under austerity measures designed to make the majority pay for a crisis they did not create – argue for curbing some of the excesses of those bosses who pay themselves multi million pound salaries on top of their multi million dollar bonuses.

The bankers and the friends of the bankers want to make everyone else pay for the crisis created by the bankers and their friends. They think that everything will then be back to normal. This is what the 1% want. Well, first of all, there never was 'normal', and second of all, it is not possible to go back.

The bankers and big business – the bosses – can act together, as a class, when they need to. This is the secret of how they remain in power. The 99% do not yet act as a block, as a class. When they do act in this way, they will, at the same time, develop the ability to consign the 1% to the dustbin of history and in turn, create the means for building a society fit for free human beings.

Capitalism has not been in existence for very long – in historical terms, it has been around for no time at all, perhaps three or four hundred years. When human beings have lived on the planet for millions of years, those among the 1% who argue that there is no alternative to the present system, Capitalism, either do not have a very good imagination or they are trying to fool us.

Part 1

Renewable Energy

Chapter 1

Ambient Energy, Intermittency and Electrical Storage

Ambient energy, or renewable energy as it more often called, is just there. We don't have to do anything to get it. Most of it originates in the sun. The sun rises every day, provides us with light and heat, warms the earth and our buildings, allows plants and trees to photosynthesise, and gives rise to wind and waves. Most of this ambient energy from the sun is either reflected or re-emitted, in a continual, complex series of energy transformations back to space, but, before this process is complete, it is possible for us to harness or trap the energy and allow it to do work for us. The energy in tidal flows and the geothermal energy from hot rocks are also forms of ambient energy.

Ambient energy sources like sunlight, wind and waves and the tides are available in a more or less predictable way all the time. A windmill can grind wheat to produce flour today and, without requiring any fuel, if the wind is blowing, it can produce flour again tomorrow and the next day and so on.

People have always attempted to control the flows of ambient energy, to harness it and make it work for them. So for example, in warm climates such as that in the Mediterranean where the heat from the sun causes wide variations in temperatures, the thermal mass of stone is used to keep buildings at a reasonable level of comfort throughout the day and night. Heavy stone buildings absorb heat from the sun during the day, keeping interiors cool. At night when outside temperatures drop, the warm stone helps to keep the house warm. The stone in these buildings is being used for heat energy storage.

Ambient energy is taken for granted most of the time, although, without it, of course, there could be no life. The question of how energy is produced, used, stored or converted to other forms of energy now, is of the greatest importance since climate scientists are reporting that the temperature of the planet is rising – to possibly dangerous levels – and that the rise is linked to the way energy is produced. The scientists have concluded that the warming is due to high and growing concentrations of greenhouse gasses in the atmosphere. The most important greenhouse gas is carbon dioxide, which is released as a by-product of producing energy and heat from the burning of fossil fuels. The Intergovernmental Panel on Climate Change[1], who have published several reports warning of the dangers, together with other concerned scientists, are in agreement that the main cause of the increased atmospheric concentrations of carbon dioxide is the burning of fuels like coal, oil and gas.

As coal, oil and gas are the world's main fuels, this predicament requires urgent attention. The objective must be to implement a safe solution as soon as possible. A rational response is to stop burning fossil fuels, replacing this with an alternative way of producing energy. But is there enough ambient energy, to provide all the power and energy needs of twenty-first century society?

There are a number of ways to convert ambient energy to electricity. One of the most viable technologies converts the energy in wind to electricity. Wind energy converters (wind turbines) which can be positioned at the favourable sites which exist in most countries are the subject of many studies attempting to determine how much energy can be harvested from the wind. In Britain for instance, a survey carried out by the UK Government Department of Trade and Industry in 2002 which attempted to evaluate the energy resource available from offshore wind, estimated that there was potential for 3213TWh[2] per year to be harvested from offshore wind turbines. No wind turbines were to be allowed within 5km of the coast for this survey. In 2007, the total energy consumption figure, including electricity, for the UK was 1860TWh[3]. So, according to those figures, the offshore wind potential is almost twice the consumption figure. Considering offshore wind energy only then and only considering sites no closer than 5km from the coast, there is more than enough harvestable ambient energy to provide all of the UK energy requirements.

A similar study in the United States by the National Renewable Energy Laboratory (NREL) has estimated the potential for wind energy to be 38,552TWh[4]. Total energy consumption in 2009 in the United States was 27,682TWh[5]. Again, then, considering only wind energy, there is more than enough harvestable renewable energy available to equal current US energy consumption. Large tracts of land such as wilderness areas, parks and urban areas are excluded from the energy potential estimates as they are considered unsuitable for wind development and so, for instance, only 1.67% of the state of California is considered available for onshore wind development, whereas, 55.54% of Texas is considered viable for wind farms.

The solar photovoltaic cell (PV) converts sunlight into electricity. A 500,000 km^2 area of desert (this is about 1.3% of the earth's total desert area) installed with PV cell technology would be able to produce electricity equivalent to total world energy consumption[6]. One study of the potential for PV technology, estimates that one third of total US electricity output could be produced by PV cells mounted on the roofs of suitable (in terms of compass orientation and absence of shading) existing domestic and commercial buildings[7].

The intermittency problem

Renewable energy sources like the wind or light from the sun are intermittent, their strength varies with time. For wind, at any one time, we can not be sure what the wind strength will be a minute or an hour later and, in places like Britain, because of the unpredictability of cloud cover, the rate at which solar energy is received at any place is famously no more predictable than wind strength is (although, of course, at night, we do know there will be no generation at all!). In other places, like, say, California, or at the equator, or in parts of the Mediterranean, the weather is a lot more predictable and so the power that can be produced from PV equipment is also more predictable – but it is still a variable power source, and one which is firmly not available for more than 50% of the time.

In order to understand the strengths of renewable energy, we must first acknowledge that, in the variability of these sources of energy, there is a challenge to be overcome, and we need to explore solutions to that challenge to discover the great advantage of renewable energy over conventional energy generation.

PV panels do not provide power at any time we want. They only provide power when the sun is shining on them. Windpower is limited in a similar way. The wind does not blow all the time or at a constant rate. When there is no wind, or not enough wind to turn the rotor blades of a wind turbine, or if the wind is too excessive for the turbine to generate power, no electricity will be generated. The opposite phenomenon is also true. If the wind is turning the blades of the wind converters, or the sun is shining on the PV panels but there is no call for electrical power, any generation of electricity is destined to be wasted.

Both these situations will cause problems for a system running entirely on renewables.

In the first case, the problem is obvious. If, at night say, all the wind turbines stop because of unsuitable wind conditions and if we are relying on PV and/or wind for electricity, then we will simply have no power – all the lights will go out and other electrical appliances will stop working. The other situation – excessive generation – is also problematical because electricity cannot simply be discarded if there is too much of it. If a suitable load is not found for the generated electricity the turbines can be stopped as a way round the problem, but this not only represents wasted energy, there is a cost to turning off wind turbines (see below). A better solution is to store the electricity for future use.

Storage

Pumped hydro electrical storage works by using any excess electrical power on the grid to pump water from a low level to a higher level lake. This creates a store of potential energy which can be converted back to electrical energy at a later time when it is needed.

There are a number of pumped hydro electrical storage stations, the largest of which, in the UK, is the pumped hydro plant at Dinorwig in Wales. The pumped hydro facilities were developed as support for large thermal power stations like coal and nuclear power. Most of the large thermal power stations are not very flexible and can not be turned up or down very quickly, and because of this they need facilities to store any excess power for times when their output is greater than demand. Any stored excess energy generation can be released from the store later when the demand has increased. Typically, the less flexible stations are run at full power, and the storage is used at night when demand drops. In this way, the system operator can arrange for the more inflexible plant to run continuously. (At least in theory; in reality, most of the UK nuclear power stations are down for maintenance most of the time[8]).

The Dinorwig pumped hydro power station in the UK cost £450M to build at 1982 prices. This reflects the fact that the storage function is very valuable for an electricity grid. This same pumped hydro technology can also be used to support intermittent renewable energy generators. However, the energy storage developed for the old nuclear and fossil fuel generators is not, as I attempt to show later, large enough to support a radical increase in the number of intermittent generators on the electricity grid. If we want to reduce and eliminate fossil fuel generation by replacing it with renewables, then it will be necessary to develop substantial storage facilities. With adequate storage, it will be possible to provide firm power with renewables. It will enable the storage of energy when there is, say, an excess of windpower on the system; and it will enable for power to be available when there is no wind and/or there is no sun for PV generation.

Renewable energy and grid operation

Electricity grids must cope with fluctuations all the time because of the continual variability of electrical demand, and so, even quite substantial numbers of wind turbines connected and delivering electricity to the grid present no additional problems to the system[9]. But there is a limit to the number of variable output generators that can be incorporated onto the system before problems are encountered.

As might be expected, the National Grid (UK) has undertaken considerable research into the effects of introducing large percentages of windpower on to its system.

It finds that the intermittency of wind does not pose a major problem for stability at penetration levels (the proportion of wind-power on the grid) of up to 10 to 20%. At penetrations above this, research into renewable energy and the grid indicates that *very substantial energy storage capacities will be required*[10] if the continued deployment of uneconomic conventional plant is to be avoided.

To some extent the characteristic unevenness in supply of wind power can be countered by having a large geographical spread of windfarms. Power delivered from a widely dispersed set of windfarms will be much less variable in nature and thus more easily managed on an electrical system than power from, say, one wind park on its own.

Currently, the system copes with short term demand changes

by having a number of grid-connected conventional fossil fuel burning generators running at half speed (half power). This is called spinning reserve capacity. The power from these kinds of generators can be ramped up or down at relatively short notice and so, they can be used to provide power when there is an unpredicted shortfall, or shutdown when there is an unplanned excess of power supply over demand. Spinning reserve operation though, is expensive and more polluting than generators working at full power because operating at half power is less efficient.

Too much windpower!

In September 2011, the UK National Grid paid £1.2million to a windfarm company because the windfarm was producing more than the National Grid could absorb.[11]

This was presumably paid as part of a contract made with the windfarm company which would have set out at the outset how much of the windpower produced would be purchased by National Grid. Similar contracts are made all the time between National Grid and 'conventional' energy generators. Payments are made to companies to 'switch on' or 'switch off' generators according to the fluctuating demand.

However, preventing wind turbines from generating electricity has a different effect to switching off a gas fired power station say. No fuel is being saved by the switch off. Energy is simply being wasted. Or more correctly, as no electricity has actually been generated, the opportunity to generate electricity from the current wind conditions is not being taken up. The maximum use is not being obtained from the wind turbines and ancillary equipment. There is a cost to lost wind energy; a cost to the system when more electricity is being generated than can be absorbed. The cost can be addressed only with storage. As touched on above, the pumped hydro storage facilities were built to reduce the cost of the thermal power stations – to reduce the wastage costs of non take up of power generation

Similarly, to ensure minimal wastage of wind or other ambient energy generation, the energy must be stored when it is abundant, and that store used for times when there is no wind or other renewable energy supply.

What kind of storage.

Just like the grid operating with conventional generators, an electrical grid, powered solely by intermittent renewable energy must be able to manage the fluctuations in supply and demand without letting the lights go out. The concept of too much or too little windpower would not exist with an electricity system equipped with adequate electrical storage facilities. If there was more electricity being produced than could be used, it would simply be stored for use at a later time when the turbines were generating less.

If we want to be able to use all the energy harnessed by a renewable energy generation system, it must be able to store energy when there is a surplus. It must have electrical storage. This is because, as discussed, due to the intermittent nature of the renewable resource (wind, sun, wave) and the variable nature of the electrical demand, there will be times when the electrical demand on the grid is less than the power being produced, while at other times the demand will be greater than that being produced.

Having enough storage is the key to being able to satisfy all our energy needs with renewables.

More storage facilities like those at Dinorwig in the UK are needed. They are expensive to build, but they will enable renewable energy sources to supply firm power.

However, there is another, arguably much better, storage solution alternative to more pumped hydro.

Vehicle to grid storage (V2G)

In following chapters in this book I explain how states like California or the UK and, by extrapolation, all states, can provide all their electricity and energy needs with renewable energy sources combined with an appropriate amount of energy storage equipment.

This appropriate energy storage could be provided by what has come to be known as Vehicle to Grid Power or V2G.

Electric Vehicles

For V2G to become a viable alternative storage system there must be a shift from the internal combustion engine (ICE) vehicle to Battery Electric Vehicles (BEV).

V2G was first described by workers at the University of Delaware in 1997.[12]

The idea behind V2G is based on the fact that for most of the time, vehicles stand idle and that while they are not being used for their primary purpose (of being driven) the expensive technology in the vehicle can be used for other purposes. The possibility of doing this comes about because the vehicle's battery not only holds the charge for driving the car forward but **it is also a valuable electricity storage device.**

The battery can be charged when there is an excess generation on the grid (perhaps because it is very windy) and, electricity can be pushed back into the grid from the vehicle battery when the power generation drops (or, in practice, when the frequency on the grid falls). The battery can be charged when the price of electricity is low (this is usually at night in the UK for instance) and with the right type of electrical connection, the battery can pump electricity back into the grid when there is a lot of load on the system and when, because of how the law of supply and demand operates currently, the electricity price is high.

In this way, a vehicle fleet connected to the grid can be of value to the electricity network system operator (National Grid in the UK, CAISO in California) as well as to the users of the vehicles.

A battery electric vehicle fleet connected into the grid could supply the same services as the existing pumped hydro schemes. Actually, because the batteries store high quality energy (electricity) which is instantaneously available, the BEVs would be able to supply better and more services to the System Operator.

Dinorwig, the UK's largest pumped hydro storage power station can hold 9100MWh of energy and can deliver 1728MW of power (for 5 hours). If, say, one million 'Smith'[13] Ampere vans, (each with a 25 kWh Lithium-ion battery) were available and plugged into the electricity system, they could store 25,000MWh energy and could deliver, with the appropriate electrical connection, 4000MW of power. (This is more than twice the storage capacity of Dinorwig and twice the power capacity). Currently there are 32 million registered (ICE) vehicles on the roads in the UK. Most of the vehicles are parked at any one time (on average people who are driving will drive for 38 minutes a day [www.statistics.gov.uk]) For the rest of the day then, (more than 23 hours) the vehicles are parked and sitting idle. If this were 32 million BEVs instead, we would have the basis for building a renewable energy generation system which would be able to produce firm renewably generated power.

It might be thought that, apart from the variation in efficiency of the various different electrical storage technologies [12], in terms of what the electrical storage on the system is introduced to accomplish, there would be no essential difference between any one form of storage and any other. In one sense, this is, on the face of it, absolutely true. However, in the context of a world which is being poisoned with greenhouse gas emissions, and because any technical and political changes required to reduce these emissions must be made quickly, it turns out that the V2G storage solution for renewable energy has not just one, but two very great advantages over all the others.

Firstly, V2G requires that internal combustion engine (ICE) vehicles are systematically replaced by battery electric vehicles (BEV). The current high cost of fuel is making electric vehicles a viable replacement for the ICE car without even considering any advantage (and it is a massive advantage) which attaches to them through using them as storage and power supply devices as well as transport which is more efficient and intrinsically less polluting. If electric vehicles are used to provide the electrical storage on the grid needed by renewables, then, for each electric vehicle introduced, there is one ICE vehicle that is not burning fossil fuels and emitting CO_2. Ultimately, these electric vehicles will replace the entire fleet of ICE vehicles and, accordingly, the entirety of the global warming gases emitted by them. Two birds with one stone! If, say, pumped hydro were to provide the storage for the electricity grid, there would not be any equivalent impact on the numbers of fossil fuel vehicles being manufactured and used.

Secondly, building enough conventional dedicated storage for the electricity grid is an extremely expensive project. Take pumped hydro for instance. Dinorwig cost half a billion UK pounds (at 1982 prices). And Dinorwig can only provide a fraction of the storage required. If, instead of ICE, electric vehicles are manufactured this will start to achieve two things, not just one. It will satisfy the continuing requirement for personal transportation *and* provide the basis for an electrical storage system that can support the switch to renewable energy generation from fossil and nuclear. The cost of building the storage is largely taken away since it comes as an integral part of the transport system.

People need transport for getting to work, college and essential trips, and, as there is no alternative at present because of an ineffective mass public transit system, if they can afford to do so, they get personal transport to achieve these aims. A large proportion of national (private plus social) investment goes into transport because vehicle manufacturers try to satisfy the existing demand for personal transport.

Because vehicles are being manufactured anyway; if, by altering the product so that it now performs two important functions, not just one, then this, arguably, is a doubling of value and productivity. And productivity which releases a second technology (renewables like wind, wave and PV) from one of its constraints (variability or intermittency).

This book argues (and suggests that campaigners raise it as a transitional demand) that all road vehicle manufacturers switch immediately to produce battery electric vehicles instead of internal combustion engine vehicles for the reasons given above. Trains, trams and other transport solutions such as streetcars (trolleybuses) can and should also be developed or converted to provide both transport and also electric storage facilities. Over time, a growing infrastructure of electric mass transit systems equipped with battery electricity storage which supports electricity generation can start to reduce the numbers of cars on the road.

In the following chapters, I show how, in detail, that, with suitable wind turbine, tidal power or PV installation and with a large enough electrical storage facility*, renewable energy can provide reliable and secure energy and electricity delivery for states like the UK and California. And by extrapolation, if it can be shown that it can be done in these two states with high energy use currently, and because solar energy is fairly evenly distributed over the earth (and coinciding with areas of high human settlement), then similar solutions can be implemented, providing that the political climate is favourable, all over the globe - in Africa, South America, Asia etc.

In the UK, enough power can be produced at times of high winds to satisfy all our electricity demands, and also, with storage, produce enough stored energy to satisfy electrical power demand for when there is little or no wind *and* satisfy all the non electricity power and energy needs.

*I estimate that a sea area slightly less than size of Wales equipped with optimally spaced very large wind turbines, with associated storage, will for instance, satisfy the entire UK requirement for energy and electricity[15] (based on the 2007 digest of UK energy statistics). Similar results can be achieved in all countries using the ambient energy sources present in those various localities.

But there are widely respected voices out there who question whether renewable energy is really a viable solution for our future energy. These voices are discussed in the next chapter, Hot Air.

Chapter 2

Hot Air

Let us not talk falsely now, the hour is getting late
Bob Dylan. All Along the Watchtower.

There are a lot of conflicting ideas put out in the press and media regarding renewable energy and, on balance, renewable energy does not get a very good press.

The following is just a short selection of the type and variety of comments that can regularly be heard on mainstream media in the UK and elsewhere:

> Wind power is an extraordinarily expensive and inefficient way of reducing CO_2 emissions when compared with the option of investing in efficient and flexible gas combined-cycle plans[1]

> I think they are of no value whatsoever[2]

> Windfarm paid £1.2 million to produce no electricity[3]

> Focusing on the role of wind power for our sustainable future makes good sense from both an environmental and an economic perspective[4]

> BP shuts alternative energy HQ[5]

> The best windfarms in the world already produce power as economically as coal, gas and nuclear generators[6]

> Offshore wind is staggeringly expensive[7]

A battlefield of ideas emerges, then, from which it could be difficult to form a sensible opinion. So, in order to get a better picture of the potential of renewable energy, people have little choice but to turn to commentators they trust. One such trusted source[8] for many people are the writings and speeches of George Monbiot, the well-known campaigner against injustices of many kinds. I have a lot of admiration for George Monbiot and his work. His book, *Heat, How to Stop the Planet Burning,* is hard hitting and well researched and raises a lot of the right questions. However, I think he has not helped the renewable energy sector by backing nuclear and, as I attempt to explain below, it is to renewables we really need to turn for an international solution to the current crises.

Another widely respected source, Professor David Mackay's book *Sustainability Without the Hot Air* was introduced to me as a good book on renewable energy. I was looking forward to agreeing with his humorous take on the energy debate. However, what appear to be substantial errors or miscalculations mean that I now question a lot of the professor's remarks.

I am going to take a critical close look at what these two influential commentators say about the prospects for renewable energy. What I read and hear often brings to mind the good sense of Karl Marx's famous motto: 'Doubt everything'[9], and unfortunately, for me, these two authors have not shown that they can be excepted from that rule.

Monbiot

> We have 25 years, ladies and gentlemen and not just to act a little bit. We have 25 years to cut carbon emission throughout our economies, the developed world economies by 90%. Now there are some people who believe this can be done by replacing the fossil fuels on which this great tottering pile has been built by ambient sources of energy. And that I'm afraid is in the realm of science fiction. Yes, there's plenty out there –

> there's plenty of wave and wind and sunlight which we're not tapping. But every year we use the stored sunlight from 400 years of accumulation, *every year 400 years worth of sunlight.'*...'Our freedoms, our comforts, our prosperity are all the result of fossil fuels. Ours are the most fortunate generations that have ever lived. Ours are the most fortunate generations that ever will.
> (My emphasis)

This is from the speech that George Monbiot gave (the gist of which was published in the Guardian[10]) in London's Grosvenor Square on one of the first Campaign Against Climate Change marches, held on 03 December 2005.[11]

To hear this was very depressing for many of us at the march.

It meant that it would not be possible for future generations to have the levels of prosperity that are currently enjoyed.

For one thing, it would be difficult to describe 'our generations' as fortunate, when, for instance, out of a population of seven billion, there are half a billion children who don't have enough food, and when two and a half million of those children die each year of malnutrition, leaving the rest with the condition of 'stunting' (meaning they never reach their full potential stature and state of health, a condition which lasts their entire lives).

But now, since hopes of a better way of producing energy are exposed as 'science fiction' by George, the conclusion of those listening to the speech (and those who read George's Guardian column[12]) could not be other than: things can only get much worse.

George says that it is necessary to build new nuclear power stations to provide the energy and power that we need. My argument with George, at this point, is not about his advocacy of nuclear, there is a big question mark over nuclear technology and I deal with it elsewhere[13]. I am concerned here with removing the question mark put over renewables in general by George's contributions.

If George is right, and we currently do use, in one year, in fossil fuel, the equivalent of 400 years' worth of the energy from the sun that reaches the earth, then, of course, it would not be possible to replace fossil fuel with ambient energy arising from sunlight. There would not be anywhere near enough. Ambient energy could only ever provide in one year, 1/400th of what we need.

Fortunately, George is not right about this. In less than one hour, more energy is delivered to the earth in the form of radiation from the sun than is used, in a whole year, by all the world's people[14]. Putting that another way: in one year, ten thousand times more energy falls on the earth from the sun than is used by modern twenty-first century capitalist society per year.

If it took four hundred years for the earth's fossil fuel energy stores to be increased by the energy equivalent of one hour of solar irradiation, then, in the same time, 3,503,999 * equivalents of one hour of solar irradiation would have been re-emitted back to space.

This is because the solar energy the earth receives is balanced[15] by an equivalent amount of energy that is lost to space.

*[400x8760 – 1] (There are 8760 hours in a year).

Only a miniscule amount of the sun's energy falling on the earth was converted into fossil fuel all those billons of years ago[16]. It could be said that the earth is not very efficient at capturing and storing the sun's energy. Most of it ends up back in outer-space.

Contrary then to George's assertion in the above quoted speech and in his writings, **there is more than enough ambient energy**. His timescale is out. To be correct he would have needed to say in the above speech: ...'*every year* [we use] ***one hour's worth of sunlight*** '... (not as he said, '**400 years**').

He is wrong by a factor of around three and a half million. And that is considering only the energy received from the sun. There is also the energy that arises from the spin and motion of the earth and its interaction with the sun and moon – that is, the tides, plus geothermal energy.

It would only be in the realm of science fiction to believe we could replace fossil fuels with ambient energy if we did not have the *means* to harness the energy. There is more than enough ambient energy several times over to replace fossil fuels. All we have to do is capture it, use it, and let it go on its way. We can not capture all the ambient energy. But there is no need to, because, as we have seen, many thousand times more energy is delivered by the sun to the earth than is currently disposed of by the earth's population as it burns fossil and nuclear fuel.

So, do we have the *means* to harness the ambient energy?

Harnessing ambient energy

Several well established, mature technologies now exist which capture ambient (or renewable) energy. The particular technology needed to harness the energy depends upon location. In Britain, offshore wind power, tidal power, wave power, solar thermal and some photo-voltaic (PV) might be one set of choices, in Spain for example: more use might be made of solar power technology and PV, together with wind technology.

Using publicly available wind, sun and tidal data, I developed a mathematical model which analyses these ambient energy sources to give some idea of the potential of renewable energy. This forms the backbone on which the book is built. In California, wind strength data and solar measurements were used to estimate how much of the technologies would be required to replace fossil fuels, whereas in Britain, wind strength data and tidal tables were used for the calculations.

Being up-beat about the possibility of renewable energy systems being able to satisfy our energy and power needs seems to divide opinion. Some of those who are not optimistic are professors of physics and energy. For example, Professor Ian Fells: 'I do not believe renewable energy will ever provide more than 20% [of electricity] (and it must be heavily subsidised)'[17]. Professor Ian Fells is not the only distinguished academic to make this kind of pronouncement on renewable energy.

Mackay

Professor David Mackay, a physicist at Cambridge University in his very influential, and much acclaimed book *Sustainable Energy Without the Hot Air*, says, regarding future energy plans, that 'we need a plan that adds up', that he is not biased towards this or that technology, but that he is 'just pro-arithmetic'[18]. Professor Mackay is not just any old professor. He is the Chief Scientific Advisor for the UK government's Department of Energy and Climate Change, so what he says and publishes is influential and, therefore important.

Details

I'm afraid I have to go into a lot of detail in order to prise some sense from Professor Mackay's arguments. There is no other way of doing it. Please bear with me. Apparently, the devil is in the detail. We have to look elsewhere for the blame though.

In the book, Professor David Mackay calculates Britain's total power consumption to be **125kWh per day per person**. (He is using power units that he developed himself throughout the book.) The professor's preferred unit for power consumption is kWh/d/p (Kilowatt-hours per day per person). (We have more energy and power units than we could possibly need already[19], why add another one?). His argument is that doing it his way allows per person comparisons to be made across countries more easily. Anyway we have to live with another unit of power in order to read the professor's book.

In the wake of an analysis of all the possible traditional

renewable energy sources he 'fears' (after a public consultation) the maximum Britain would ever get from renewables is in the ballpark of **18kWh per day per person**. In other words he says he fears Britain could only get about 15% of its energy from renewables.

Both of these professors reach similar conclusions then. How depressing (again).

After reading through Professor Mackay's book for a second time, a few things, to me, didn't seem to add up. This was despite the fact that 'this is a book about numbers' – with constant reminders in the text about the importance of this, like; '...numbers not adjectives', 'With numbers in place, we will be better placed...', 'This heated debate is fundamentally about numbers', 'It's about physical limits to sustainable energy, not current economic feasibility', 'This is a straight-talking book about the numbers' and so on. If you don't have Professor Mackay's book to hand, you might want to skip to the next section until you have it. (Even though I am disagreeing with many aspects of Prof. Mackay's book it is a very good starting point for discussions on energy. And it can be downloaded free[20].)

The first time I noticed when something didn't add up was when he calculated a figure of 40kWh per day per person for car consumption and compared this with a figure for maximum plausible power production from onshore 'windmills' of 20kWh per day per person. The UK car consumption is **twice** the plausible maximum for UK onshore wind[21]. This conclusion is worrying if true. But, is it right? To answer this requires only a little analysis. So... he is comparing consumption due to car diving with the maximum plausible production from onshore wind turbines.

Firstly the consumption figure for cars is worked out and turns out to be as mentioned above, **40kWh/d**[22]. (There is a 'per' missing off the end of this unit in the professor's text. This allows for a wrong figure to be slipped into the table).

The 40kWh/d figure (on page 29 of the professor's book) is actually referring to 40kWh per day *per car driver.*

The figure is entered into the kWh/p/d (kilowatt hours per day *per person*) consumption column. **It should not go into that column**.

The unit for that number (40) is **not** kilowatt hours per *person* per day (kWh/p/d), but, kilowatt hours per *driver* per day (kWh/d/d). [Apologies for introducing yet another unit as if we haven't got enough energy units already]. The two units are not the same thing because there are 59.5 million persons, but only around 30 million drivers.

Putting that consumption per *driver* figure into the consumption per *person* column is a surprising schoolchild-like error because the **consumption** column is being compared with plausible energy **production** from onshore renewables per *person* column. This calculation assumes that everyone in the country drives 50 km in a car every day. This is obviously not the case. The UK population of around 60 million do not all drive, let alone drive a car 50 kms every day. The 40kWh per day is the averaged consumption for a typical *driver* not the consumption quota for *one member of the population.* (Even this is not very accurate, because the average UK daily driving distance is around 37 km, not the 50 km which is being used[23]).

So, if the driver consumption figure is 40kWh per day per *driver,* then, the **correct consumption for car driving per day per** ***person*** figure is: 40kWh/day x 30M/60M. (***20kWh per day per person*** not ***40kWh per day per person***).

Professor Mackay's conclusion about the relationship between UK car consumption and maximum plausible energy production from onshore wind is simply wrong.

We can now correct Professor Mackay's conclusion:

The UK car consumption is ***the same as*** the professor's calculated figure (see below) for the plausible maximum for UK onshore wind (***not twice as much***). ***Professor Mackay is wrong by a factor of 2.***

The erroneous first estimate for car consumption remains in place throughout the professor's book. That on its own is enough to ask Professor Mackay to check his sums and to say sorry for misleading us. For a book all about the numbers, it is almost laughable. Since car energy consumption is one of the bigger energy expenditures the error is not a laughing matter – an inflation of 100% for the car consumption figure is not trivial.

Actually, the professor may have been a little embarrassed by his sums, because, at the end of the chapter he offers what is supposed to be an explanation of the figures. Unfortunately, this explanation simply reinforces the problems with the calculated results. This explanatory note (for page 29) is presumably included in order to back up the assertion that the car distance traveled per day per person in the UK is **50km**.

I looked at the UK government statistics for motor vehicle traffic volume for 2009.[24]

This is given as 313.2 billion vehicle miles (504 billion vehicle kilometres). Using the figure for UK population that Professor Mackay uses (59.5 million), gives a UK average for vehicle distance traveled per day per person of **23km** (not the 50km that the professor uses). Actually, 23km is an over-estimate because the UK government statistics figure for motor vehicle traffic volume for 2009 given above includes lorry and bus mileage and is not simply one for cars. The figure for vehicle distance traveled per day per person then, is actually **less than half** of what was used by Mackay in the calculations he used to estimate consumption. This corroborates my previous calculations which show that the professor's calculations are wrong by a factor of more than two.

Professor Mackay explains that he is basing his calculations on a typical moderately-affluent person and that this is valid since most people aspire to reach that level of affluence.

It may be thought to be commendable to include such aspirations, given the current levels of inequality present with us in twenty-first century capitalism, but really, having the whole population driving, in calculations which also make a claim to objectivity for comparisons of energy consumption versus maximum plausible renewable energy generation, only serves, because it is not based on actual consumption, to introduce doubt into the mathematics of something which needs to be discussed in a straightforward manner.

The professor then, again, is a factor of more than 2 out. This is a very significant error. And you have to start to ask why these glaring errors are included in what purports to be a book about the numbers.

Onshore wind

Moving now to the professor's calculation for the maximum plausible UK energy production from onshore wind; as alluded to above, the professor calculates that the maximum plausible power we can produce from onshore wind is 20kWh/p/d.

The plausible wind figure is calculated and 'plopped' (one of Prof. Mackay's favourite words) into the column adjacent to the consumption column. The maximum plausible production from onshore windfarms is arrived at by assuming a typical onshore windspeed of 6m/s. This windspeed corresponds to a power per square metre of land surface area figure – of $2W/m^2$. By **guessing!** a figure of 10% for the proportion of the country to be covered with wind turbines, the professor deduces a figure of **20kWh/d/p** for the plausible generation from onshore wind . A figure is plucked out of the air for the fraction of the country to be covered with 'windmills' (by the way, he knows they are not windmills, why does he call them that?) And where did the 10% come from? Why not 5%, or 20%? Choosing 5% would halve the maximum plausible production, choosing 20% would double it. To base the maximum plausible production from onshore wind turbines in the United Kingdom, on what is obviously simply a guess, in a book about numbers, is, to use another of the professor's favourite words, codswallop.

Politicians, or bankers, argue in this way, plucking the number 10 randomly, rather than 5, or 20, to prove a point; raise or lower their bonuses etc. They get away with it all the time. Physicists, though, are better than that at numbers, aren't they?

Mackay's conclusion

The conclusion according to Professor Mackay is that "Britain's onshore energy may be 'huge', but it's evidently not as huge as our huge consumption". We have already seen how the estimate for consumption is an over-estimate. Is the plausible production figure being estimated correctly?

Professor Mackay answers this himself by emphasising that the assumptions he is making are generous. In other words he is saying that his generous plausible maximum production figure (20kWh/d/p) for onshore wind is an *over-estimate* and that the actual figure will be less than that. He illustrates this argument with comments like:

'the windmills that would be required to provide the UK with 20kWh/day per person would amount to 50 times the entire wind-hardware of Denmark; 7 times all the wind farms of Germany: and double the entire fleet of all wind turbines in the world'.

In a fledgling industry like the wind turbine industry, it should come as no surprise that there are not very many of them *anywhere* in the world. Arguing like this is a bit like a farmer in 1769 saying to James Watt 'Replacing horses with the steam engine? You must be mad, there are millions of horses around the world and there is only one steam engine'.

Our conclusion from the professor's calculations must be that he, the DECC chief scientific advisor, has calculated that there is not enough energy plausibly available from harnessing onshore wind in the UK to match the UK consumption, not that they will take up lots of space, although the professor claims he is 'simply trying to convey that helpful fact… that the wind farms must cover a very large area'.

Professor Mackay is arguing firstly, that the maximum average onshore power per unit area that can plausibly be produced from wind is $2W/m^2$, and that, secondly, this is not enough to match our 'huge' energy consumption[25]. The figure arrived at by the professor for the 'average onshore power per unit area' (**$2W/m^2$**) is highly significant, and one that I think needs to checked carefully.

To verify the figure of $2W/m^2$ the professor calculates the power per unit area that will be available on a specific real wind farm (Whitelee) being developed near Glasgow.

This wind farm is specified as having 140 turbines with a combined peak capacity of 322MW in an area of $55km^2$. Unfortunately, no real wind speed figures are used to do this calculation, but nevertheless, as we shall see, he arrives at the exact same figure – ($2W/m^2$) – for the maximum average onshore power per unit area that can plausibly be produced from onshore wind for the real wind farm, as the figure he arrived at by plucking the figure of 10% out of thin air.

The professor has usefully explained how to calculate the power per unit area that can be harnessed from a wind farm[26]. The equations give us the maximum packing density for turbines in a wind farm[27]. Not surprisingly, the optimum number of turbines you can cram into a wind farm of a specific area depends upon the rotor blade length of the individual turbines. You can get a lot of small turbines in or fewer larger ones.

We are told that the Whitelee wind farm is to have 140 turbines in an area of 55 square kilometers. Professor Mackay does not tell us about the specifications (rotor blade length or rated power) for the individual wind turbines. However, using the figures we are given, we can work that out.

Using the equations[28] given, we can calculate the optimum blade length for 140 turbines crammed into 55 square kilometers to be 62.5m. We do not know anything about the wind regime at the wind farm, but we can assume that the designers of the wind farm envisaged wind speeds exceeding the average, because real wind speeds are never at the average but are always above or below it. In fact in parts of the UK, such as on hills or on flat plains, as we all know, it can get very windy indeed (that is of course why wind farms are always only built on land in exposed areas – they are only installed in the areas which get high winds). So lets assume, the wind farm designers assumed that wind speeds at turbine height could reach speeds of up to 15m/s. Yes, I am doing exactly what I complained the professor was doing, plucking numbers out of thin air. Yes I am, but please bear with me it will come right at the end! Here, I am using a plausible windspeed temporarily, to help me make my point. I will go back to using real evidence before the argument is finished. (Actually, a lot of wind turbines are built to withstand gusts of up to 25m/s, at which speeds, they are shut down to save the equipment). At 15m/s, a wind turbine with a rotor length of 62.5 could produce around 8MW depending on the actual turbine design. So for 140 turbines, with the optimum rotor arm length of 62.5m, the peak power (achieved at wind speeds of 15m/s) for the wind farm is greater than 1GW (8 x 140).

The peak power produced by the wind farm, then, is over 1000MW.

As the wind farm is occupying an area of 55km^2, we can calculate that the peak power per unit area figure is over 18W/m^2 (1000/55). The peak power reached per unit area for the wind farm is over**18Wm^2**. This is *more than three times* the peak power figure arrived at by the professor (He arrives at **6Wm^2**). Perhaps I am making mistakes. I don't think so. I invite you to go over the figures and check them for yourselves.

Applying to the peak value, the same load factor of 33% that the professor uses (the load factor is a value which is applied to the peak capacity of a wind turbine to take into account the fact that the power generated varies in strength depending on wind patterns at that site), gives us a value for average power production per unit land area (at Whitelee) of ***6W/m²***.

This is *three* times that important figure (of **2*W/m²***) arrived at by Professor Mackay.

Having checked my figures again, the question arises: Why is this?

There are four factors.

Firstly Professor Mackay does not talk about real wind speeds. He talks of the average, or typical wind speed for the whole country. The reality is that some areas on the planet have low average wind speeds, and others have higher average wind speeds. Wind farms are built in areas which have higher than the average wind speeds and thus higher than average power production per unit land area. The average power production per unit land area at the sites where wind farms are likely to be located is greater than $2W/m^2$. This is an almost self-evident factor.

Secondly, we do not really know enough about the design constraints for the Whitelee wind farm. We could hazard a guess that they may have planned to install larger turbines at a later date (when turbines of that size became available) and installed smaller turbines initially. Most onshore wind turbines currently have a much shorter blade length than 62.5m simply because turbines of that size are not yet (in 2014) commonly being installed or manufactured. If that was the plan, it would make sense, because, the initial electrical connections could be designed to also cope with the larger currents produced by the bigger turbines that might be connected later.

There may have been financial reasons why less capacity was being installed than was physically and technically possible. From the information available, we cannot know.

What we do know is, *less capacity is being installed* in the 55 square kilometres than could have been, and this is why the average predicted power production per unit land area comes out at only $2W/m^2$ and not $6W/m^2$.

We can work out that the capacity of the turbines actually being installed at Whitelee is 2.3MW (322/140). Given that less capacity is being installed than the optimum, how much land would 140 2.3MW turbines occupy at optimum spacing? The type of generator and rotor designed to be installed at any particular site depends on the predicted wind regime and we do not know this detail so, again, I'll have to make some assumptions here. Making assumptions for receipt of winds of, say, again, up to 15m/s, the design rotor length for a 2.3MW turbine would be, depending on the efficiency factor, in the region of 45m [see Appendix]. If installed according to the optimum spacing equation given in Mackay, these would take up an area of about $30km^2$. This leaves an area of *25km² empty of wind turbines* on the Whitelee wind farm.

So, if the topography was suitable, the wind farm operator had enough land area to have installed around three hundred optimally spaced 2.3MW turbines on the $55km^2$ wind farm area. That is more than twice the number actually being installed (One hundred and forty were planned to be installed). If the extra turbines were installed on the site, the W/m^2 figure, the whole reason for this exercise, would be considerably higher than Professor Mackay's $2W/m^2$.

As briefly referred to above, Professor Mackay's book is about the *physical limits* to sustainable energy. Whatever is limiting the power per square metre figure in the case of Whitelee wind farm, it is not a physical limitation relating to wind strength. The limitation in this case, as is shown below, is a social, economic or political one.

The third factor is the most important one. Real wind speeds are much more important than average wind speeds. This is because the power produced by a wind turbine is proportional to the *cube* of the wind speed[29].

Take an average wind speed of 6m/s. If, over three hours, the wind at a particular place was roughly constant at around 6m/s, the average wind speed would be 6m/s. The power produced in the first hour, would be proportional to 216 (6^3), and in the second and third hour it would be the same (proportional to 216). Over these three hours, the energy produced would be proportional to 648.

If, over the next three hours, the wind speed in the fourth and fifth hour was zero (ie no wind at all) and in the sixth hour, it was 18m/s, say, the average wind speed would be still be 6m/s. Over the fourth and fifth hours the energy produced would be zero. In the sixth hour it would be proportional to 5832 (18^3). The energy produced is *nine* times more in the second three hours than the first three hour period, even though the average wind speed is the same in both.

Wind farm developers can use average wind speeds as a guide to selecting the best sites, but they can not use the average wind speed across a country to work out the available energy resource; they need to have a good idea of the actual wind speeds at particular locations. They measure wind speeds over a number of years and calculate precisely where the best places are for the wind farms. Even if the average wind speed at a given site was the same as the national average, it would not mean that the average power density there was the same as the national average. As we have seen, it can be considerably larger, or, of course, smaller.

What is important is not so much the power per unit land or sea area, but the actual power that can be generated given the equipment available and the characteristics of the wind at the specific site. W *not* W/m^2! The generators must be sited where they will generate the most power. It doesn't really matter how much space is needed. Actually the area that wind farms take up is fairly irrelevant, since the land can continue to be used, in many cases, for exactly what it was used for before. Sheep are quite happy grazing on wind farms for instance.

And finally, I decided to a bit of research on the Whitelee wind farm.

It turns out that the actual rotor blade length to be used at the site for the 140 wind turbines is, in fact, the same as the number derived above, 45m[30]. Applying this value in the equation[31] for calculating the optimum land area for 140 turbines shows that these turbines would occupy a space of around 30 square kilometers. Since Whitelee wind farm occupies an area of 55 km^2, that leaves an area of 25 km^2 unused!

A little bit more research throws up this interesting fact:

'Whitelee is also home to a 25 square kilometer area of habitat management, one of the largest in the UK with lots of interesting species to spot like merlin and black grouse' [32]. Could it be that Whitelee wind farm proper only takes up an area of 30 km^2?

If so, the professor's calculations for power per unit area are simply wrong since his calculations are based on 55km^2 as being the wind farm size.

To start to wind matters up about power per unit area for wind farms, I want to take a look at a second real wind farm to, hopefully, further disabuse people of the notion that our onshore wind energy resource is not as huge as our consumption.

The owners of the Carnedd Wen Wind Farm site near Machynlleth in Wales plan to install 150 3MW turbines. The map for the wind turbines shows them to be spaced 400 metres apart. Each turbine then, occupies 160,000m^2. If each turbines peak capacity is 3MW, the peak power per unit area is 18.75W/m^2. Applying the load factor of 33%, the average power production per unit area is **6W/m^2** [33].

As Professor Mackay arrived at a much lower figure, (**2W/m^2**) a question mark must be placed over the conclusions he has made for the maximum plausible power production for UK onshore wind.

The professor has over-estimated the energy consumption and under-estimated the maximum plausible energy production figures from onshore wind.

As the Chief Scientific Advisor for the UK government's Department of Energy and Climate Change, what Professor Mackay says and publishes is influential and, therefore important.

If I had known at the time of reading his book, that Professor Mackay was not just any old professor, but that he was the Chief Scientific Advisor for the UK Government's Department of Energy and Climate Change (DECC), I might not have been so surprised by the anomalies recorded above. The professor also does work for the World Economic Forum (WEF). This is an organisation where the rich and powerful from across the globe meet to discuss their vision of the world and attempt to shape the future world in their image. For the huge multinational oil, gas and coal companies and for the nuclear power industry, wind power must present as an unwanted competitor.

Offshore wind

There are similar problems with the professor's estimate for plausible offshore wind energy production. Mainly, the problems are that the offshore figures are based on the onshore figures. As the onshore estimates are underestimates, this means that the offshore estimates will also be wrong. There is also an attempt to argue that there is not enough sea area available. If that is the case, we really are in trouble. We do have plenty of sea. If there are constraints, they are not physical as such, but political or economic. Yes there will be problems with the type of seabed, depth of water, problems with salt in the atmosphere and corrosion etc. These can all be overcome with proper forward planning and using well understood marine technology.

I could spend the whole of this book picking apart the

professor's conclusions, but I do not want to burden the reader with any more of this as I feel I have already exposed more than enough questionable detail in the calculations to serve my purpose.

What is my purpose? To show how we can provide all of our energy needs from ambient energy sources. Professor Mackay is correct about the fact that the wind turbines will cover a large area. But the area is not the problem.

First of all, the best wind resource is found where people do not live or work – out at sea. There is no shortage of square miles of open sea. There is more than enough shallow water, for producing at sea, all the power needed[34]. And there are a number of other good reasons for siting wind farms at sea. Marine wind speeds tend to be higher and steadier than onshore wind speeds and as the power obtainable from a turbine is proportional to the cube of the wind speed, significantly more power can be obtained at sea. Also, wind speeds increase with altitude and reaching these higher speeds necessarily means larger wind turbines. Since there are fewer obstacles to siting very large structures at sea rather than on land, siting at sea is arguably often the best proposition for maximising renewable energy production.

In discussions about wind power, it is quite common to hear people saying things like 'There is a problem with wind',[35] 'Wind is not reliable. Look out of the window, the leaves are not moving. You can't close industry down if the wind is not blowing'[36].

What these commentators are usually referring to is the variable or intermittent nature of wind. The variability or intermittency problem was discussed in the first chapter.

The next chapter, 'Powering the World with Renewables' using real data for windspeed, tidal range and solar insolation, shows how ambient energy sources together with energy storage, can, by *working with* the variability of renewable energy, deliver all the energy requirements and more for all the people on the planet.

Chapter 3

Powering the World with Renewables

In this chapter I want to show how the energy needs of all of humanity can be met by renewable sources. This isn't really such a surprising challenge, since for the vast majority of the time that human beings have existed, their energy needs have been met by renewables. It is only in the last few hundred years that carbon, embedded in the earth's crust, has, first as coal, then as oil and gas, been excavated and burned to produce energy.

To show how the goal can be reached, I will show how two high energy consumption states – California and the UK can satisfy all of their energy requirements with renewables, and indeed, be net exporters of clean energy. Other parts of the world have similar or lower requirements than these two developed states and will receive a similar amount of energy from the sun and so, my argument goes, if the UK and California can be powered with renewables, then anywhere and everywhere can.

Meeting the challenge isn't really the problem though; it is the timescale that we need to do it in. We cannot take decades to start moving back to renewables if we want to avoid the dangers of climate change[1].

Starting with the UK

In 2007, the total UK 'final energy consumption' figure[2] was 1856 TWh[3]. Most of this energy was produced by burning fossil fuels with the balance being made up by electricity produced in nuclear power stations (2.7%) and renewable energy (3.3%). I plan to show how these figures can be overhauled so that, over twenty years, the respective energy figures will be Coal, Oil, Gas (0%), Nuclear (0%), Renewables (100%). The main result of this will be drastically reduced CO_2 emissions.

There is no doubt that achieving this as an actual fact on the ground will require an enormous amount of planning and work. However, this is a good thing, since work of this nature will help to remove the scourge of unemployment that humiliates workers in practically every country.

Instead of bailing out failed bankers, and lining the pockets of wealthy investors, governments can be investing in a clean renewable future and transforming the lives of millions of workers in the process.

Offshore Wind

The UK is situated in one of the windiest parts of Europe.

How windy?

I have obtained wind speed measurements taken from a number of buoys at various locations off the coast of Britain. Taking just one of the buoys for now and picking a day and time almost at random – the 8th of February 2007 at eleven in the morning – the anemometer on buoy number 303 was reading 17 knots (8.7m/s)[4] .Noting that a lot of assumptions and approximations will need to be made in the answer, an interesting hypothetical question is: What size of wind farm would have been needed at location 303 to satisfy the entire power demand from the UK at eleven in the morning on the 8th of February 2007?

As mentioned above, the total UK energy consumption in 2007 was 1856TWh.

Assuming that the power demand in the UK is constant, (In fact, the power demand is not constant, it varies constantly) initially, in order to get a very rough idea of the size of the problem, we can say the continuous power demand of the UK is 1856/8760 = 0.211TW[5]. The wind farm must be capable of delivering 211GW of power.

Calculations[6] show that around 23,000 wind turbines with a blade length of 67 metres will deliver 211GW from a wind speed of 8.7m/s [making reasonable assumptions about power losses in the cables and equipment[7], and for how wind speed alters with height and sea state etc.]. Optimal spacing of these turbines means that the wind farm would take up an area of sea of around 10,000 square kilometres. That is an area equivalent to half the size of Wales.

If we could rely on steady demand and a steady wind speed of 8.7m/s for an offshore wind farm half the size of Wales, then, *all of the UK's energy requirement* (electricity, transport, heating etc.) could be satisfied by that single massive wind farm. But, and here's the rub, demand and wind speeds are not constant.

One hour later on the 8th February 2007, at buoy 303, the measured wind speed had dropped from 8.75m/s to 8.23m/s and this means that the hypothetical wind farm would not have been able to satisfy the total power requirement at that time. Even worse, at 11pm on the same day, the wind speed had dropped below 2m/s. As this wind speed is below the speed at which most large wind turbines are designed to operate, *no electricity at all* would have been generated at that time.

Electrical energy storage

Clearly, commissioning one single massive wind farm to power the whole UK is not a sensible plan. Before moving on though, I want to bring electrical energy storage into the discussion.

Earlier on the 8th of February 2007 (at 10am), the anemometer reading for buoy 303 was 19 knots (9.8m/s). This wind speed meant that the power output from the wind farm would have been considerably *more* than the UK power requirement. The power output at 10am would have been about 246GW or 35GW greater than the demand. With appropriate storage facilities, 35 GWh of energy could have been saved over that hour for use later (for example, at 12am when the wind speed had dropped again (to 16 knots). In fact, without storage, or another way of dealing with the excess generation, having 35GW of surplus power (equivalent to about 35 coal fired or nuclear power stations) is extremely problematical and costly. The storage needed to balance the differences in power output arising from the above wind speed change from 19 to 17 knots is equivalent to the storage capacity of three of the UK's largest energy storage facilities – three Dinorwigs[8]. Of course, a sensible strategy for offshore wind would be to not locate all of the turbines in the same sea area. The turbines should be as widely dispersed as possible in order to reduce the correlation between the turbines, to reduce the variability of output and achieve the maximum generation stability.

Even so, even with a widely dispersed suite of wind farms, there will always be variability in the magnitude of power produced. This means that sometimes the power supply will be more than the power demand and vice versa – sometimes the power generated from the totality of wind turbines will, because of wind conditions, not be enough to meet the immediate demand. If renewable energy is to be a primary source of power, energy storage will always be an essential requirement[9]. This is because, apart from biomass and perhaps geothermal energy, practically all the renewable generators – tidal range, tidal stream, wave power and solar PV – are variable sources of supply.

There are several storage technologies that may be capable of delivering the magnitude of energy storage capacity required. A large number of pumped hydro facilities for instance could in principal deliver the required amount. The problem of pumped hydro is the high cost and the problem of finding suitable sites; although with focussed investment and political will it is of course achievable. The facility must be capable not only of storing energy, but of delivering enough power for the system requirements. For instance, Dinorwig can store about 9100MWh, but it is not capable of delivering 9100MW back to the grid (ie. delivering back all the stored energy in one hour). Its maximum power output is 1728MW. This is because of the specification of its installed turbines.

Hydrogen

Electrolysis of water to produce hydrogen which is burnt at a later time in a power station or processed in a fuel cell to produce electricity may turn out to be a good way of storing energy, however, there is more development time needed before the technology can be fully available. The hydrogen storage cycle is not very efficient. More than twice as much energy is needed to convert the water to hydrogen and oxygen and back again to water than is produced as usable energy. Pumped hydro is much more efficient (75%) than hydrogen storage (30%).

Another storage technology currently being developed is so called cryo-power where air is chilled to a liquid state using surplus electricity. The liquid air is then allowed to warm to drive a turbine to produce electricity when it is needed. Claims have been made of 70% efficiency for this technology. However efficiencies of 70% can only be achieved when the system is linked to a thermal power station (like nuclear or coal fired, or bio-mass): the exhaust heat from the power station is used to warm the liquid air, making both the power generation and the regenerated power cycle more efficient. On its own (ie. working with renewable energy generation like wind power), cryo-power is only 25% efficient.

Using batteries to store and re-use energy has been proven to be more than 80% efficient[10]. This high efficiency is one reason why I think that batteries and, specifically, electric vehicle batteries, will play a large role in storing future renewably generated electricity.

Why electric vehicle batteries?

First of all I am assuming that electric vehicles will start to become the predominant form of transport, at least in the near future. I assume this for a number of reasons. In the second decade of the twenty-first century, the price of fuel is a major concern for motorists – the price of oil and gas has gone through the roof, and electric vehicles are becoming increasingly competitive in terms of fuelling costs. Electric vehicles can be around 5 times[11] as efficient as traditional gas/petrol vehicles because electric motors are more efficient in their use of energy than the internal combustion engine – the ICE wastes up to 75% of its fuel intake as heat. And, also, electric vehicles have a number of other distinct advantages over ICE vehicles[12]. These advantages, combined with the specific advantage I want to explore here – vehicle to grid storage and power[13] to support renewable energy generation – make electric vehicles very attractive.

Vehicle to grid

What follows is a quick 'back of the envelope' example of what a fleet of 32,500 million (around the size of the current UK ICE vehicle fleet) distributed typical electric vans connected to the UK grid might be able to facilitate.

One typical commercial vehicle – the Smith Electric Vehicles Ampere – has a battery which can store 24kWh – enough energy for an average household for a couple of days. 32,500 million of them have a combined storage of 780GWh. This is about the same amount of storage as 86 Dinorwig pumped hydro stations.

The total energy demand in the UK in 2007 was 1856TWh. The electricity demand was 354TWh. A fully charged electric vehicle fleet, then, having previously stored surplus renewable energy, could 'keep the lights on' in the UK all by itself for about 19 hours even if there was no wind anywhere around the UK. And, for the purposes of this hypothetical scenario, no extra special electrical connections between vehicle and grid would be needed since the electrical current in the vehicle to grid connection would be 5 amps[14] – well within the UK domestic household limit.

Studies for the UK have shown that the likelihood of long

periods of time when there is no wind at any point anywhere around the UK at all is very low[15]. And, over time, a power system which is predominantly renewables-based could, with sufficient storage available, accrue year by year, greater and greater stores of surplus dispatchable (instantly deployable) energy.

In this 'back of the envelope' scenario, the fully charged vehicle battery fleet could keep the equivalent of the UK's electrical equipment (using 2007 figures) running for 19 hours. It could supply the UK total energy requirement for 3.7 hours.

These are theoretical amounts of time and in practice, of course, the whole of the UK will not be run entirely from the vehicle fleet. I have put these figures in illustratively. They simply attempt to show the potential for a contribution to the grid from a fleet of battery electric vehicles. So, the fully charged fleet could provide the power equivalent of 221 nuclear power stations running at full power for 3.7 hours providing all our industrial, heating, electrical, and transport energy requirements for that amount of time. Yes, theory and practice are very different – but it is necessary to have a theory – without one you cannot create good practice!

In practice, nothing is static. The output from the renewable generators varies and the demand for power varies – from minute to minute. The vehicle fleet storage size is finite. You could imagine how, if the demand was low for a period and/or the renewable output was greater than demand for an extended period, that the fleet battery would be fully 'topped up'. When this condition is reached, any more surplus energy would start to be wasted (and cause problems because you cannot simply 'throw away' electricity if there is no demand for it).

Model

I set up a model[16] to see how, using a fleet of electric vehicles as storage, how many wind turbines would be needed and what size and type of storage would be needed to a) provide all the electricity for the UK for a particular year and all the energy for the UK transport for that year (with the transport fleet running on electricity not fossil fuel), and b) meet the total UK energy consumption figure. To do this I acquired some real wind speed data[17] for UK coastal waters. I used this to estimate from hour to hour over a whole year what power could be produced and synchronised the result alongside the UK electricity demand for the same period. It was difficult to get exactly the data I needed so I have had to 'massage' the data a little. I wanted to create a model which showed some degree of dispersion of the wind farms (ie. not having all the turbines in exactly the same sea area). I could only get data for three locations from 2007. But I had data from 2002 for four. The way round was to consider that the seven wind patterns I had were from the same year i.e. 2007 and match that to the power demand data from 2007. This is not as good as having 7 sets of meteorological data for the same year, but I think it is good enough to show what I want.

The equations and tables in the appendix give an idea of how I used the raw meteorological data to show how power can be developed from renewable resources to match the UK energy requirement. Initially, the model only matches the electricity demand. However, because of the way that electrical vehicles are being used in this model, matching the electricity demand also means that the transport demand is met at the same time. As might be anticipated as a result of this, the amount of electricity on the UK grid increases as a result – the electricity demand goes up. This is to be expected. What I am arguing for here is that fossil fuel consumption is replaced by consumption of renewable energy – renewable energy in the form of electricity. There will need to be a degree of what is called 'grid strengthening' – ensuring that the existing network is capable of carrying the increased loads resulting from the increased demand from electric vehicles and from the electrical storage systems aboard mass transit systems. However, the amount of increased demand on the grid from transport is considerably lower than the current amount of energy consumed by transport. This is because of the greater efficiency of electric transport compared to ICE transport (diesel and petrol motors of all types). Additionally, because of the distributed nature of the connections from the grid to the electric storage and demand manifested by electric vehicles and other electric transport systems, the grid strengthening of the existing cables etc. can be carried out gradually and in tandem as the ICE fleet is gradually banished by the electric fleet.

More jobs! Yes, and that is good.

Eventually, the remaining energy demand (heating and the industrial energy demands) is included in the model. This demand is met by direct renewable generation, the accumulated electrical storage, and, when necessary, by the stored surplus renewable energy accumulated over days, weeks, months and years. (In fact this was not necessary in the scenario illustrated in the tables in the appendix).

A calculation is made for each of the hours in the year (2007) to determine whether there is a deficit or surplus of energy when the available renewable power is matched against the total demand for that hour. Any surplus is added to the energy storage value and any deficit is made up, when possible, by extracting power from the energy storage.

Table 1 (Daily Energy Balance) in the appendix shows how this process works by looking at 24 hours in one day in 2007.

I have attempted to provide a twenty-year timeline[18] illustrating how we can get from here to there! This is of course only a rough outline. The modelling indicates that, without additional energy efficiency measures, V2G storage and power can be enough to balance the UK electricity (and transport) demand using current typical vehicle battery capacities (eg 25kWh for a Ford Transit Connect for example), but that this level of storage is, on its own, insufficient to satisfy the storage requirements for the whole of the UK at current energy demand levels. The model shows that there is a trade off between the amount of storage, the amount of wind farm dispersion and the total number of turbines. Security of supply can be achieved with various different permutations of storage size and number of turbines and with more or less dispersion of the wind farms. Generally, more wind turbines installed means less storage is required, and vice versa, and the more dispersion there is, the less storage is needed (and vice versa). It is likely that other storage and storage to power technologies will be needed (such as H_2 storage combined with H_2 power generation or dedicated static flow batteries or static hydrogen fuel cells, pumped hydro storage etc.) unless the electrical storage capacity of a standard vehicle can be increased (by about tenfold) in the intervening period.

My model indicates, however, that with additional energy efficiency measures that lead to an overall reduction in demand by about 45%, additional storage (non V2G storage) is *not* required. This, naturally, needs verification by more work, but it is good enough for me now in showing the possibilities for renewable energy supported by electrical storage. There are other strategies that can achieve useful results (for example electrical demand management) but these are well beyond the scope of this book.

As Zero Carbon Britain[19] and other studies[20] indicate, maximising energy efficiency is a necessary part of any strategy towards moving to a zero carbon world. I think that is right, and believe that demand reduction measures such as, insulating all buildings to Passivhaus[21] standards, and maximising the efficiency of all appliances is essential in the task of attempting to replace fossil energy with a clean safe alternative. More jobs! Good.

Actually, the UK demand for energy fell from the 2007 figure of 1856TWh to 1660TWh in 2011. I have used the lower 2011 figure as the target to reach. The fall was mainly due to the recession, and partly due to warmer weather.

In 2011, UK energy was coming from: Coal, Oil and Gas (92.5%), Nuclear (3.5%) and Renewables (4%). The 2011 target figure, therefore, for renewables like wind and tide, can be reduced to 96/100 x 1660TWh = 1594TWh (the current nuclear and fossil fuel share). With appropriate energy efficiency measures, that figure can be reduced considerably. Besides adopting state of the art energy efficiency measures, there are other ways of reducing demand while at the same time maintaining or improving living standards. The amount of waste produced by twenty-first century capitalism is phenomenal[22]. Removal of the unnecessary production of waste (duplication of effort, over-packaging, throw-away commodities, built in redundancy etc.) will be an important part of any sensible strategy. Some careful studies[23] have indicated that a 50% reduction in energy use from current levels should be achievable through the application of a variety of efficiency measures and I have used that figure initially as a starting point for the amount of 'Powerdown' in the model. It is important to point out here that a significant proportion of the energy reductions will come about purely because electric transport is much more efficient than the fossil fuel transport it replaces.

Of course, a society with priorities based on need, not profit, one with priorities very different to the present one, would be able to easily reduce demand further. For instance, resource use and energy expenditure by the armaments and military is a significant waste that needs removal. A lot of duplication of effort and the consequent energy use occurs as a result of the competition between firms and countries in the search for ever more profit.

Errors

With work being carried out on developing more energy efficient appliances, then, and on insulating and updating buildings to the highest standards (eg passivhaus), and on reducing the unnecessary production of waste, in tandem with the work on developing new renewable infrastructure, the twenty-year energy target for renewables to deliver is, 50% of 1594TWh or 797TWh. I have assumed that there will be a continuously developing year by year increase in the level of energy efficiency achieved, resulting in 40TWh less demand per year in each successive year, and a final energy demand of around 800TWh by year twenty.

In fact, with the parameters I used, the model indicates that all carbon emissions can be eliminated half way through year nineteen, with only a 45% reduction from the current demand. Of course although I have tried to eliminate them, the model I have constructed may have errors in it, but I hope and believe that there are no serious (wrong by a factor of 2 or more) errors. I think it gives an indication of what is achievable. It was my aim to counter the pessimists regarding the potential for renewable energy. I wanted to show what is technically possible, to show what *could* be done and also, to show very roughly, an overview of how it could be done. I think what I have done with the model achieves this.

Timeline

What the timeline attempts to illustrate is that, in less than 20 years, with a 45% reduction in energy consumption, a system based primarily on offshore wind power, with 44 thousand 10MW wind turbines supported by 1360GWh of electrical storage will be capable of supporting all the functionality delivered now almost entirely by fossil fuels.

No step change in scientific know how is required – all the components are commercially available now.

I am not arguing that electric vehicles *must* supply the storage, but what I *am* arguing is that storage is an essential for variable output generators like wind. Given that storage is an essential component, to me, it makes sense to investigate the possibility of using the potential symbiotic relationship between electric transport and renewable power.

Storage of renewable energy in the form of hydrogen, which is then mixed with gas from biomass to produce synthetic gas which is then burned in existing gas power stations is being put forward as a solution in the third (2013) Zero Carbon Britain report (Allen, 2013). This sounds like, and is, a good technical solution, but the use of biomass for energy, as discussed in the next chapter, will lead to competition for land which is currently used for harvesting food and this will tend to force the price of food up. So the question is, is it a good economic and political solution? There is also a limit on the amount of energy that can be produced from biomass in the UK. According to the Offshore Valuation Report, (Helweg-Larsen, 2010) 'The government has estimated that an area of farmland equal to 25% of the UK would only provide enough bio-energy resources to meet 8-12% of the country's energy needs'.

Some people seem to argue that cars are the problem. Cars are dangerous, noisy, unhealthy, and when you find yourself on foot at the entrance to the Blackwall Tunnel, horrible – yes they can be. But they are not *the* problem, any more than pushbikes are the solution. Cars are part of the problem for current society and pushbikes are part of the solution. I think that electric vehicles and electrified mass transit systems (equipped with electrical storage – rail and train to grid!, tram (streetcar) to grid!, trolleybus to grid! are part of the solution alongside a greater take up of walking and cycling – but I think that roads must be much more safe before putting forward that people en masse take their life in their hands by cycling on today's urban and rural roads.

Six a day

I put the tidal range component in as a way of emphasising that we need to use a variety of renewable sources – this will lead to a smoothed out power demand curve – with less variation from hour to hour which in turn will mean less storage requirement. However as I attempted to illustrate (in Table 3 in the appendix), the tidal barrage, whilst delivering a reliable and predictable output once up and running, will take many years to actually build. Each individual wind turbine takes much less time from the design stage to commissioning. Once running, they immediately start contributing to carbon dioxide emissions reduction (see eg Gross et al 2006).

The timeline attempts to show how the UK can get all its *energy* from renewables. What is not shown in the timeline is that, after only *eight* years of building turbines and electric vehicles at the rate required to complete that job in less than twenty years, the equivalent of the power consumed by the entire *electricity* grid and the fleet of electric vehicles (11.8 million of them at that stage) can be carbon free. This becomes possible because electric vehicles, as emphasised previously, are several times more energy efficient than the internal combustion engine alternative, and because the storage contained within the fleet of vehicles has allowed its partnering fleet of wind farms to contribute most (not all, because some is still wasted because of lack of enough storage) of the energy harnessed by the turbines.

The offshore wind turbines must be installed at a rate of an average of 6 each and every day for a total of eighteen years. A huge task! Well, that is the scale of the problem. My timeline, as discussed, also assumes that there is an average reduction in all energy consumption of 45%. Without that energy efficiency contribution, the task is a whole lot bigger.

When at full speed, this project will be producing and installing 10 offshore wind turbines each and every day.

National Climate Service

How is this huge task going to be accomplished?

The task is an enormous one, but it is one which people will be excited at being involved with and one which they will be proud of being part of. I think the task should be pursued along the lines which a number of UK trades' unions have set out in a recently produced pamphlet[24]. They argue for the setting up of a National Climate Service to oversee and deliver what is needed. Most importantly, this project requires millions of jobs. One of the first jobs of an NCS will be to nip in the bud, the incipient fracking (Hydraulic fracturing) industry. Fracking is not needed, is reckless towards the environment in many ways, will lead to increased greenhouse gas emissions, will not deliver anywhere near as many jobs as a properly funded renewables industry and will simply continue the tired old game of the fossil fuel companies which continue to scour the earth for oil, gas and coal profit and which treat the planet and a lot of the people on it with contempt.

The alternative is to plan for a safer environment.

Ten cities

For what it's worth, for the UK (similar plans can be conceived for all other parts of the world), what I think should be done is something like this:

Identify ten cities which will be tasked with producing and installing up to one offshore wind turbine or other renewable generator equivalent each and every day for the next ten to twenty years. These cities and tasks could be something like this:

Glasgow – one offshore wind turbine manufactured and delivered every day.

Aberdeen – develop and implement tidal stream technology. Aim to install 10MW per day

Edinburgh – beat Glasgow. A healthy friendly competition between cities to meet targets might help to speed up the tempo of delivery, and time is not on our side.

Belfast – One 10MW offshore wind turbine every day plus

develop tidal stream technology.

Newcastle – One 10MW offshore wind turbine per day.

Dublin – Links must be established between all neighbouring countries. One 10MW offshore wind turbine per day plus develop wave power technology and aim to implement 10MW per day.

Bristol – One 10MW offshore wind turbine per day plus develop either tidal stream technology (10 MW per day) or work, with Cardiff, on developing the Severn barrage project (eventually this must produce up to around 4GW).

Cardiff – One 10MW offshore wind turbine per day and work with Bristol on the Severn barrage project

Swansea – One 10MW offshore wind turbine per day.

Plymouth – One 10MW offshore wind turbine per day plus work on developing geothermal technology.

London – One 10MW offshore wind turbine per day plus develop solar thermal, solar PV and ground and air sourced heat pump technology.

Ipswich – One 10MW offshore wind turbine per day.

Norwich – beat Ipswich!

Portsmouth - Design and manufacture the specialist ships that will be required for marine renewables installations.

The NCS would co-ordinate all the work and would bring together teams to carry out systematic work on bringing (retro fitting) all buildings (domestic and industrial) infrastructure up to Passivhaus standards.

Electric

For this all to work, a lot, if not all or most, of the things that we do now by burning things (gas central heating, driving cars) will in future be done with electricity.

This will not be so bad. A lot of stuff is already electric – lighting, communications, myriads of useful devices, washing machines, fridges, vacuum cleaners, all kinds of things driven by electric motors.

The vehicles we use now will have to change. This is also not a bad thing. Ride a pushbike – that is good for you. But if we do opt for having some electric vehicles, we will find, I think, that they are so much better than the ICE vehicles they replace:

No more exhaust fumes.

No more replacing worn out exhausts.

No more oil changes or oily engines.

No noisy internal combustion engine!

No need for carburettors or fuel injection systems.

No more need to buy expensive petrol.

Etc.

With very few individual moving parts, electric motors are more reliable and robust than internal combustion engines. Electric vehicles are cheaper to run, can be charged from home or place of work, are much quieter than the alternative, produce no fumes which can lead to better air quality and healthier city dwellers. It will be possible I think, in the future, to integrate the electric vehicle fleet with mass transit systems. And they can support renewable energy generators by providing storage for a vehicle to grid system!

Providing the electricity they use is provided by renewable energy generators, electric vehicles do not contribute towards CO_2 emmissions. Of course, currently, electricity is produced by burning fossil fuel, so the claim that EVs are cleaner than the ICE car they may replace may not be true if, say, coal fired power stations are producing the electricity. The in use energy cost of EVs is less than the equivalent ICE car because, as has been remarked, EVs are more efficient. However, the energy cost of manufacture is higher for at least some EVs. So, the sooner is a switch made from coal to wind, the sooner can carbon dioxide emissions be brought down.

Evaluation of the economic cost of the kind of work discussed here is well beyond the scope of this book. However The

Offshore Valuation (Helweg-Larsen, 2010) examines the potential for offshore renewables. They came up with infra-structure costs of between £200B and £900B over 40 years. (As the wealth of the 500,000 'high net worth individuals' in the UK has increased by £123B in each of the last three years, it can be seen that that sort of expenditure is easily within reach).

I won't here describe the detailed electrical connections and technical aspects of V2G since this is set out much better than I could elsewhere[25].

In the above, I have described a way of providing an emissions-free UK energy system.

I wanted to show that it can be done because many voices argue otherwise – as was stated earlier.

California

Renewable energy can provide the basis for energy infrastructures at all points on the planet. The sun's energy falls everywhere that people have settled. Electricity can be generated directly from the sun using photovoltaic technology or solar power technology or it can be generated using the effects of the sun – wind, wave etc. The actual technology deployed at any location may vary, but the arguments I used above apply everywhere. The amount of storage required will be different when, say, PV technology is used to collect direct solar energy, as opposed to when wind is the main driver, because of their differing time-related energy density patterns. Solar insolation changes strength based on a 12 hour pattern, whereas wind strength varies from minute to minute. To help illustrate this, I developed a model similar to the one discussed above, but this time based on Californian energy demand and the potential Californian renewable resource.

The wind resource of California is not as good as that of the UK. However, this deficit is balanced by the advantage California has over the UK in terms of sunny days. California, the sunshine state, has a much better solar resource than Britain, and will be able to generate a good deal of its energy requirement from solar energy converters. In order to see how much could be generated, and how much equipment would be needed, I looked at a scenario which combined energy generated from four hypothetical marine wind farms together with a varying number of photovoltaic (PV) panels. I adopted a slightly different and simpler approach in building this model.

The Californian coast

I assumed there were four massive wind farms off the Californian coast at locations where I knew the real hourly wind speeds for all the hours in 2007, and I worked out, as for the UK, the power output that each wind farm could provide. Alongside these figures, I lined up the hypothetical output from PV panels using real insolation data from one site in California over all the hours in 2007. By varying the number of PV panels and the number of wind turbines (combined with electrical storage mounted in the electric transport), and by balancing this against the known total electricity consumption for California, I came up with some rough numbers for the area of PV, number of wind turbines and number of electric vehicles (the amount of storage) that would be required to provide all of California's *electricity* (combined with the electrical energy needed for transport) in 2007.

Due to the nature of renewable energy, in satisfying this demand, there will tend to be a generation surplus which, when captured and stored, can be used to satisfy the remaining demand. This surplus is used to satisfy the remaining demand.

The UK could, in theory, produce all its energy from one source – offshore wind. The smaller wind resource in California alongside its almost permanent daytime supply of sunlight means that an efficient implementation strategy for renewable generation here will follow this naturally occurring division of resource.

If everybody in California slept when it was dark and only used energy during the daytime when the sun was shining, then it might be easier to build an energy delivery infrastructure reliant solely upon direct solar insolation. But of course, we live in a 24-hour society, and rely on electricity being available around the clock.

From the publicly available figures[26], then, for energy demand (2001) and vehicle use (2007) in California, I calculated, using a model similar to the one developed for the UK, that in one scenario example, 27,200 10MW offshore wind turbines, 100 square kilometres of PV panels (18% efficient) and the storage provided by 29 million vehicles could satisfy the electricity demand and transport energy demand for California. The surplus energy that could not be captured by the available vehicle storage in that scenario amounted to 574TWh. It turns out that that surplus provides twice the balance required in California for: heating, industrial, jet fuel, and all other uses.

However this is a mammoth task. The target is in the order of 14,000 panels per day for 20 years. The target for wind turbines is just as challenging. But the challenge is not off the scale of achievability. For example, the GE 1.5MW wind turbine weighs 164 tons. Roughly that equates to about 100 tons per MW. The total weight of the Californian wind turbine target is, using this rough guide, 27 million tons. More than this tonnage of steel and modern technology is produced by the UK auto industry every fifteen years or so. Similar scale challenges were faced and met in WW2. For example, in less than 4 years, almost 3000 Liberty ships each containing 7000 tons of steel were built in American shipyards (Mackay 2009, p62) The energy and climate crisis could be compared to facing the threat of war. The economy needs to be orientated accordingly – not to fight a war, but to build a better and safer world.

As in the UK, with appropriate energy efficiency measures, (in the region of 50%), it would be possible to satisfy all the energy demand of California with renewable generation of electricity combined with storage from electric transport.

People in love with jet air transport will perhaps scoff at the following, but, because bio-fuels are not, in my opinion, part of a sensible strategy (food has to come first, at least until basic human needs are satisfied), air transport requires some radical re-thinking and redesign. Perhaps helium or hydrogen-filled lighter than air vehicles (air-ships) will provide more leisurely long distance travel in the future. I don't know. Battery powered planes are not currently feasible. Actually, more than enough food is produced now for all. It's just that the priorities of a capitalist system mean that a lot go without and many die of malnutrition. In a rational society, where people's needs come first, there would in fact be land available to produce non-food products such as bio-fuel for jets if it was needed. However, under the capitalist system, where profits trump needs, producing bio-fuels simply pushes up the price of food and puts it beyond the reach of the poorest, because food and fuel will compete on the same market.

Why twenty years?

Basically because science is telling us we need to get a move on.

For example, a couple of months ago, I went to a meeting in London presented by Professor Wadhams of Cambridge University. Every year, for several decades, he has been measuring the thickness of the Arctic ice. He predicts that the Arctic Sea will be ice-free in the summer by 2015/16. The Arctic ice, according to his measurements, has been getting thinner year by year. Not all scientists agree with his conclusions on timing. However, what the vast majority of scientists now agree on is that the planet is warming and that the warming is caused by human activity – the activity of burning fossil fuels. A growing number of scientists are also now beginning to say that the number and strength of some of the more extreme recent weather events – such as 2012's Hurricane Sandy and 2013's exceptional typhoons in Colorado – can be linked to human induced global warming[27]. What is not contested, except perhaps by a handful of oil heads, is that, if the Arctic Sea does become ice-free in the summer, the rate of planetary warming will increase. This is because the ice helps to keep the planet cool by reflecting the sun's radiation back to space. If the ice disappears in summer, this reflective surface will be replaced by a surface which will absorb the heat from the sun.

Gordon Brown, the then UK Prime Minister, three years before Hurricane Sandy hit, said of the Copenhagen Climate Conference held that year that that meeting was the 'last chance' to reach international agreement on greenhouse gas reductions. No agreement was reached. Half a decade on, politicians of every stripe seem to have dropped any ideas they might have had on tackling the threat of global warming and the concomitant climate disruption.

I listened to all three of the 2012 US presidential television debates. The words 'Climate Change' did not pass the lips of either candidate.

After the election, President Obama has talked of 'Green Energy' and of making vehicles more fuel efficient, but even after the experience of the East Coast of America being battered by the enormous Hurricane Sandy, and the loss of human life associated with it, the US President is apparently reluctant to link disruption of the climate to the burning of fossil fuels.

What I have done so far and in the main part of this chapter was just part of the task I set myself in the introduction.

Politics

The main task is not technical, it is political.

One problem is – why isn't anything vaguely like what I have attempted to describe being carried out? In fact, contrary to this, companies are scrambling to get the dirtiest oil and gas out of the ground – from the tar sands of Canada and from the shale rocks of America, the UK and other places, and from deep beneath the surface and the surface of the oceans at all points of the globe.

It is not really the oil they are after. What they want from the oil is profit.

Governments, their militaries, arms companies, oil companies and various other agencies of government and big corporations combine in an unholy alliance to achieve their common objective: securing, by any means necessary, a steady flow of profit.

Tens of thousands of people have been killed in wars fought for oil and oil profits in the last decade.

Why? And is there an alternative?

The next chapters attempt to provide some of the answers to these political questions.

Part 2

Workers' Power

Chapter 4

The Enemy Within

So, renewable power can replace fossil power.

Why is this important?

There are at least three reasons.

It is important because burning fossil fuel is contributing to the 'Greenhouse Effect' and. consequently, is causing global warming.

Briefly, this refers to the way that adding more and more carbon dioxide to the atmosphere adds to the already existing carbon dioxide 'blanket' that surrounds the planet. This thickening blanket prevents heat escaping from the planet which, in turn, results in an increase to the earth's average surface temperature.

It is also important because of reasons of security of energy supply. Producing energy from renewables as described in the previous chapters with appropriate technology to store any surplus will provide the basis for a dependable secure supply of power delivered over the electricity grid.

Another reason it is important is price.

Although a concerted effort is being made by some interested parties to say that power from offshore wind turbines and other renewables is 'the most expensive way of producing electricity', in fact the opposite is likely to be true. For example, The Confederation of British Industry (CBI) recently published a report, 'Decision Time' (now, no longer available on their website) which raised concerns that 'low marginal cost' energy (the kind produced by wind farms) 'may drive low or negative prices'[1]. That CBI report says that the price of energy will come down when it is produced by renewables instead of mainly by fossil fuels.

If burning fossil fuels is causing a problem, and there is a carbon dioxide-free alternative, it should be easy to solve the problem shouldn't it? Phase out fossil fuel, and replace it with renewables. If that was done, we would start to see emissions fall, we would have a secure supply of energy and prices would stabilize.

But energy prices are sky-rocketing, and, looking to the future, there is no sense that the current regimes will deliver secure energy supplies. And what is actually happening to carbon dioxide emissions? They are continuing to rise year on year, because, every year, more fossil fuel is burned than in the preceding year.

In 2011, CO_2 emissions from fossil fuels burning and cement production increased by 3%. A total of 34.7 billion tons of CO_2 was emitted to the atmosphere. These emissions were the highest in human history and 54% higher than in 1990 (the Kyoto Protocol[2] reference year). In 2011, coal burning was responsible for 43% of the total emissions, oil 34%, gas 18%, and cement 5%[3]

Measurements of the atmospheric CO_2 concentration carried out since 1958 on the Mauna Loa observatory and elsewhere show that the concentration of CO_2 in the atmosphere is steadily rising. Recent CO_2 concentration measurements in parts per million (ppm) are:

2010 – 389.92,

2011 – 391.65,

2012 – 393.84,

2013 – > 400[4]

Why are emissions still rising when scientists and technicians in the field know how to stop the rise? It is because the fossil fuel problem is not a technical one that can be solved purely by technical changes. The problem is deeply political, economic and social.

The fossil fuel industry is not only at the root of the problem of climate change, it is also at the heart of the capitalist system itself. A vast web of concerns linked to the profits from the oil, gas and coal industries are dug in and provide support for the continuing extraction of oil, gas and coal from the bowels of the earth. Oil companies, banks, coal corporations, gas specialists, arms manufacturers, auto manufacturers, petroleum refining industries, chemical industries, pharmaceutical industries, the electricity generation industry; all these and many more have a vested interest in the continuance of the fossil fuel industry and the profits it generates.

Fossil fuel has been a two-edged sword though. It has creative as well as destructive qualities. The energy that is and has been produced by the fossil fuel industry has allowed for a much more rapid expansion of the world economy since the industrial revolution of the eighteenth and nineteenth centuries than would have occurred had a fuel like coal or oil not been discovered.

Without the energy that is available, now, from fossil fuel, it will be almost impossible to build a renewable energy system from scratch. Try building a 10MW wind-turbine with human or horse power. Given the scale of the energy challenge ahead, the continuance of the fossil fuel industry for a *planned* number of years will be a necessary pre-requisite for the development of a viable renewable energy alternative to that fossil energy system.

But the changes to the system of energy production must be planned. As renewable power comes on line, existing energy producing stock will be retired. And this is where the problem comes in.

Because, far from planning the retirement of fossil power, each and every day, capital and energy is invested in new fossil technology plant, in new fossil fuel research and development and in searching for more deposits of the stuff etc.

A lot of the existing energy related capital has very long economic lifetimes. As far as the owners of this capital is concerned, oil, coal and gas can and should continue to be mined, sold and burned, decades into the future – well past any rational date determined by the science of climate change.

To comply with any properly planned schedule of renewable energy infrastructure installation, some fossil fuel related stock will have to be retired 'early'. This will mean that the oil companies and the banks will have to do without their expected 'returns' – their pound of flesh, their profits – on their investments. Any challenge or threat to the profits flowing from oil will impinge severely on several key industries and any such threat, therefore, will be and is being resisted[5]. This resistance will be and is from the heart of the system, from the heart of the beast.

Some evidence of the centrality of the oil, coal and gas industries can be seen by taking a look at a list of the world's most profitable companies. Top of the list in 2012 was Gazprom, the Russian fossil fuel giant with profits of $44.5B. Next on the list is Exxon-Mobil, the US based multi-national with $41B. Then comes a bank – The Industrial and Commercial Bank of China with profits of $32B. This is followed by Royal Dutch Shell (number 1 on the Fortune 500 list) with a mere $31B. Positions 5, 8, 9 and 11 are filled by oil companies, number 6 is another bank and number 10 is The Ford Motor Company. (The odd one out at no 7 is Apple).

Of course, the car companies in their present form could not survive without the fossil fuel industry to produce the stuff we put in our cars and other vehicles. And vice versa, the fossil fuel industry must have a market for its primary product – oil. The two industries are mutually dependent and, no doubt, chief executives of each other's companies sit on each other's respective decision making boards with the goal of maximising profits for both industries[6]. The banks make loans to both industries with investments for building refineries, retooling of factories etc., stretching repayment schedules far into the future. So there is a web of interests backing the fossil fuel industry and the ICE car manufacturing industry to continue making profits far into the future. For the fossil fuel and associated companies, anything that upsets this 'Business as usual scenario' is an unwanted interference and a threat to their profits.

Climate change, or at least efforts to combat it, is one of these threats; and this can be seen from the various campaigns by the oil industry and its friends to sow doubt in the public mind about the science of climate change[7]. This confusion sown by the so-called 'climate skeptics' helps Royal Dutch Shell, for example, and its competitors plough ahead with their gamble of drilling for oil in the Arctic and the Mexican Gulf, fracking for shale gas in Cumbria and Sussex and dredging the surface of the land in Canada for tar sands etc.

The web of interests referred to, include, perhaps, not all capitalist interests or 'capitals', but, because all capitals benefit from a share of the total pool of profit, (the profit made by one capitalist gets shared out, not in a planned way, but it gets distributed via the way the market operates – eg. with banks, landlords, retailers, legal services, arms manufacturers, other services, governments, all getting a cut of the profits) they act in support of one another (that is when they are not trying to wipe each other out through competition) – they act as a class to defend their interests – the capitalist class interests. The oil companies and their friends are at the core of the capitalist system and strive to ensure it works in their interest – in the interests of the capitalist class, the ruling class.

There are two main components of the crisis as I see it - the crisis in the capitalist economy and the environmental crisis. The crises have the same cause as I attempt to argue below. What is needed though is a solution. I attempt to sketch a way out in the next chapter. The current chapter now focuses on how we have arrived at this mess and how it will be difficult for the current 'business as usual' system to reverse out of it without major social change.

The double outrage is:

1) It is totally acceptable and 'business as usual' for political leaderships, that, while billions are robbed of the essentials needed for a healthy happy life, a handful of capitalists conspire with government authorities to cut ever more deeply into the already blighted lives of the working poor they live off, in order to take ever more wealth for themselves. These cuts are made in many ways. Fishermen on the Mississippi Delta whose catch was decimated because of the BP oil pollution pay; children and poor families working in sweatshop industries, making items that they can never afford to own pay – sometime with their lives. The injustices carried out in the search for profit are unmistakable and happen all over the globe.

2) That while these capitalists are busy feathering their own nests, the ecosystem that they take for granted and on which all life depends is slowly being killed off by their actions.

There is however hope amongst the horror and it can be found in the resistance that is sprouting and springing up all around the globe in response to the real injustices of the system. I don't pretend that I even get near to having all the answers; this is a very complicated world. But it seems to me there are a few things that need asserting:

1) Capitalists are reckless in their search for ever more profits.
2) The blind drive for profits by capitalists always leads to unintended consequences for the capitalists themselves and for the rest of us.
3) The political leaderships worldwide are part of the capitalist class, are part of the problem, and are in a deadly embrace with the big corporations and their networks of support, and, like rabbits in the headlights, are unable to do anything to avoid the coming catastrophe.

The above three factors are part of a whole and are interrelated but I have separated them out as far as I can.

1) Recklessness

As I tried to show in the previous chapters, it is very possible technically to reduce to zero the environmentally damaging emissions of carbon dioxide that result from the burning of fossil fuels. This reduction, however, is not happening and as we have seen, the emissions continue to increase. Why is this?

Capitalism treats the environment as its dustbin – and it has always done so.

For an example, on the night of 2nd-3rd December 1984, in the city of Bhopal, in the Indian state of Madhya Pradesh, a gas leaking from a pesticide manufacturing plant owned by a household name billion dollar multinational chemicals company, immediately killed thousands of people as they slept in their homes. Shortly afterwards, thousands more died as they tried to escape. Estimates vary on the exact figures for how many lost their lives as a result of the leak, but, according to the Guardian[8], more than 15,000 people died and more than half a million suffered injuries.

No workers were killed on the site that night. The day workers and their families, asleep in their shanty town huts next to the factory, were not so lucky. The gas (methyl iso-cyanate) rolled into their rooms and killed them as they laid there. Almost 30 years later, the plant has still not been cleaned up[9]. The poison remains as an insult and a danger to every citizen of Bhopal. Why did it happen? I think the culprit is profit.

The primary motive of all capitalists is profit. Because profit is the motive, because it is the primary reason for the production of any commodity, everything else, as a logical consequence, takes *second place*. First place goes to profit. That profit is the profit of the few, at the expense of the many, at the expense of the environment. The safety and health of the many and of the environment are not of prime concern – by definition. In fact they seem to be of no concern at all judging from the actions of many of the world's major corporations.

On 10th April 2010, the BP oil and gas rig, Deepwater Horizon, exploded. Eleven men, who were working on the platform, were killed by the explosion. Seventeen other workers were injured and millions of barrels of polluting crude oil were pumped into the Mexican Gulf. Industrial disasters are not one-off incidents; they are part and parcel of 'business as usual'. The so called 'accidents' at work kill and maim workers on a daily basis. Major disasters like Bhopal and the BP Gulf of Mexico outrage make the headlines. Other incidents, perhaps because they are deemed as less news-worthy, happen regularly and, largely, go unreported. We only get to hear about some of them.

On 5th April 2010, five days before the BP explosion in the Gulf, another 'accident' at work, this time an explosion at the Upper Big Branch Mine in Montcoal, West Virginia claimed the lives of 29 coal miners. It was suspected that a build up of methane and inflammable coal dust led to the blast. The company running the mine had a history of violations (against regulations) and had been fined for not putting adequate ventilation and safety measures in place after a similar fatal incident four years earlier at another of its mines when 12 miners lost their lives in an explosion. New safety regulations were put in place after the 2006 explosion, but they were, effectively, ignored. After the 2010 incident operations simply carried on, business as usual. It is never the CEOs of drilling, mining and fracking[10] companies that get killed in these incidents, but always the workers who pay the ultimate price for the pursuit of profit for the few. These are examples but, if you look, you can find an almost never ending litany of accidents and killings at work.

The pattern seems to be: After an 'accident', there is an investigation by the state authorities, new regulations are put in place, the company is fined and then carries on in the same old way, with much wringing of hands, but with 'business as usual' until the next 'accident', when the company is fined again and so on. In other words, the state says to the company you can carry on business as usual, but if you kill someone or pollute the environment and get found out, there will be a *financial* penalty to pay. Occasionally there is a fall guy whose fingers get severely wrapped with a prison sentence, but this is rare and often it is the workers themselves who find themselves the scapegoat.

This is not a problem restricted to 'The West'. Reports of accidents at work originate from all parts of the globe. The most deadly coal mines on the planet are in China. There are, on average, 13 deaths a day there from fires, floods and explosions. Fifty thousand coal miners have lost their lives in China over the past decade[11].

'Accidents' at work are an inevitable consequence of the priorities and working conditions that people are forced to accept. Workers are treated as an expendable resource, as potential 'collateral damage' in the pursuit of profit. The environment is plundered for resources and used as landfill.

BP and the other fossil fuel companies are now racing to establish drilling operations in the Arctic. They know they are taking a risk. But it is not a risk that they will pay the price for. No one wants to see an incident like the Deepwater Horizon happen in the Arctic. Not even the oil companies want to see this. But that won't stop them. It is inevitable that more oil disasters will happen and will happen in the Arctic, given the operational history of the world's fossil fuel companies. It is inevitable unless they are stopped. The companies themselves *will not stop*. After all they are only doing what good capitalists do, aren't they? They take risks. And like all good entrepreneurs they reap the rewards for taking those risks, don't they? Their governments demand they do this. Whether it's the American government, Russian government, British government, Chinese government or French government etc., these states want to see a steady stream of taxes and revenue from investments that are made in these vast projects. After all, the British and American governments recently attempted to bomb a country back to the Stone Age in order to gain control of the profits from oil, why would they baulk at yet another oil spill or a few more dead workers? The reason America invests so much with its allies in the Middle East is because the region sits on a half of the earth's known oil reserves and the American state with its associated oil companies and other supporters wants to ensure that the profits from this reserve flow towards America and its allies around the globe. Incidentally, because of the centrality of oil to the capitalist system, it also dominates lives in less obvious ways. So, for example, it can be argued that, if there was no demand for oil, the Palestinians would be free[12].

I think the evidence shows that the oil companies and their networks of support are reckless in the pursuit of profit. The policies being pursued by the big fossil corporations, the banks that finance them and the states which support them, and the risks that are being taken with people's lives and people's livelihoods and with the air that we breathe and the water we drink and even the ground we cultivate and stand on, are not leading any of us to a safe place. And now, by ignoring the understanding and the science relating to climate change, about how global warming is linked to greenhouse gas emissions, these capitalists are taking a new, much greater, gamble than has ever been taken before.

A multinational company took risks with its Bhopal plant and that led to the release of a toxic gas that killed thousands and blighted the lives of hundreds of thousands of innocent people. That company simply paid a fine and walked away. Based on risk and cost benefit analysis, the company would have decided that it was worth taking the risk based on the likely amount of the fine. The chemicals company reportedly[13] paid, in an out of court settlement, around $470 million dollars to the Madhya

Pradesh state.

The world's large corporations and their respective states do pay lip service to the need for controls on greenhouse gas emissions. They see 'the market' as the way to control the emissions. They talk about 'the price of carbon', carbon permits and of carbon 'offsets' as part of the way towards achieving carbon dioxide emissions reductions. Trading in 'carbon' is now a multi-billion dollar industry. People are invited to put money into carbon credit schemes (some of which can turn out to be dodgy scams) with the promise of high returns on investment. The theory is that an increasing 'price of carbon' will lead to a reduction in carbon dioxide emissions as firms and people seek a less costly energy alternative to burning fossil fuels through switching to renewables or being more energy efficient .

But this reliance on the market is not bringing down the rate of carbon emissions – they continue to rise. It seems as if the carbon price is being set so that it will inflict no financial pain or only minor pain on the major fossil fuel corporations.

The carbon dioxide emissions produced from the burning of fossil fuels can be seen as a deadly gas escaping into the atmosphere 'as we sleep' that will, as at Bhopal, kill people, only more slowly, or blight their lives. If effective measures are not adopted leading to a rapid lowering of the concentration of the gas in the atmosphere, the authorities could be charged with the offence of reckless criminal negligence, of taking a risk with, not thousands of lives, but with all of our lives, by gambling with the health and safety of the planet.

The actions of capitalists can cause misery for workers. The capitalists do not, of course, plan this. But in taking risks to maximize profit, they plan to risk causing misery for their workers. In Bangladesh this year, more than a thousand workers died and thousands more lost limbs and suffered traumatic injuries when the factory building they were working in collapsed upon them without warning. The building collapsed because it was cheaper for the capitalist to let people work in a decaying dangerous structure built on a swamp than it was to provide a safe working environment. The basic logic within capitalism means it is certain that tragedies like this will keep occurring and repeating over and over again until some fundamental, systemic, social changes occur.

Left to itself, capital will continue to crush the life out of more and more workers and, eventually the life, as we know it, of the planet. This is so because, in effect, the capitalist is blind to the misery he causes, for the capitalist, profit trumps all. It trumps the life of its workers and it will trump the concerns of any environmental activist who pits rational arguments for ecological safety against the interests of capital. This will continue to be the case even if some individual capitalists regret the damage being inflicted on people and planet - they may wring their hands, but the misery continues. And now, as the global temperature rises, the capitalists are guilty, as some scientists are desperately pointing out, of planning to risk irreversible damage to the planet's eco-systems.

Any solution to the climate crisis, to the problem of increasing levels of greenhouse gases in the atmosphere, the problem of potential catastrophic climate disruption, must be an international one. This is because any gas escaping into the atmosphere at any point on the surface of the globe will be quickly dispersed and will eventually spread fairly evenly through the entire atmosphere. It is necessary therefore to drastically reduce all emissions of greenhouse gases like carbon dioxide wherever they occur. We have seen how it is possible to achieve these reductions with solutions incorporating energy efficiency and renewable energy generation.

It is also possible and necessary to now build flood defences in places like the low-lying plains of Bangladesh where the factory workers were killed. Areas like this at many points on the globe will be at risk of flooding due to the climate disruption which the scientists are telling us is going to continue, even if emissions were to stop now.

But neither the reductions nor the building are happening. The conclusion can only be, again, that capitalism's capitalists are recklessly negligent. They will not stop this, they do not have the motive. Therefore, we must use all the power that we have to stop them.

2) The blind drive for profit

In this section I attempt to point to the tendencies within the capitalist system that always lead to crisis and to also show how the renewables industry fits, in this context, within a capitalist system dominated by the fossil fuel industry.

To start with then; Can capitalism defuse the crisis it has created for itself, of ever increasing CO_2 emissions and the resultant global warming? In one sense, capitalism has *already failed* that task. It has been known for several decades that the continuous increase of CO_2 emissions will lead to dangerous climate change. The droughts and floods that have been occurring around the globe, at an apparently increasing frequency, are being ascribed by some climate scientists to global warming. People are losing their lives, and livelihoods, due to extreme weather events which are more and more being attributed to the climate change arising as a result of human activity. And yet, far from reversing the level of greenhouse gas emissions, current policy measures[14] ensure atmospheric concentrations of greenhouse gases will continue to increase faster than ever before. So, why are the big corporations and governments acting as they are?

I think that if the oil, gas and coal corporations could, without endangering their profits, switch to building the renewable energy and energy efficiency alternatives to fossil fuel then they would waste no more time and they would make that technical switch as quickly as they could.

The trouble for them, though, is the cost of that switch. That is, the cost to those corporations, not the cost to the rest of us. Vast amounts of capital are tied up in the fossil energy industry, so switching lock, stock and barrel to renewables would mean there would be significant loss incurred by major capitalists in the form of stranded investments. The result is they try to put off any change. Also, a lot of influential technologists and scientists employed by the fossil fuel industry see their careers linked to the continuance of oil etc. This contributes to the substantial inertia and intransigence within the industry. Turning round the energy industry is, in more ways than one, like trying to turn round an oil tanker.

The price of energy

There are some capitalists investing heavily in renewables, but, at present, the scale of this investment is puny compared to the investment still going into new fossil infrastructure, nevertheless, the fossil fuel industry is threatened in a unique way by renewables.

This threat comes about as a result of a fatal contradiction built in to the heart of capital. Capitalism is the most dynamic system humans have ever known. More 'stuff' has been produced since the inception of this system a mere three or four centuries ago than was produced in all of preceding history. And yet, capitalism hates a surplus. If there is a 'glut', an over supply of something, the price drops and it becomes very hard to sell.

The price of energy, of course, is fundamental when it comes to discussions about energy.

If you believe what a lot of mainstream commentators are saying, then you would believe that 'Offshore wind is the most expensive way of producing electricity.' And that if a switch is made to renewables, then the price of energy will rise. But is that right? The premise of the statement is partly correct. But read correctly it means that, *for the fossil fuel companies*, offshore wind (and we can also add onshore wind and other renewables) is the most expensive way of producing electricity. What is the explanation for this?

Each MWh of electricity generated by a wind turbine is 1MWh of electricity not sold by the coal, oil, gas or nuclear energy companies. Each MWh of renewable electricity sold on the open market is 1MWh that the fossil fuel and nuclear sectors would prefer that *they* sold. But it is what happens to the *price* of energy when that energy is produced by renewables that is really worrying energy companies. The price of any commodity (transport, housing, food, energy, any widget etc.) is partly controlled by the laws of supply and demand[14]. If there is 'too much supply' of any particular commodity on the market, the price of that commodity will fall as a result. The Organisation for Petroleum Exporting Countries (OPEC) knows this law well and uses it to raise or lower the price of oil when it sees fit by curtailing or increasing, respectively, the amount of oil its members place on the World market.

When, in 2011, the Fukushima disaster in Japan occurred, and Japan switched off all its nuclear power, there was a danger that the price of energy would soar as Japan turned to gas for its energy to replace the missing nuclear reactor output. The price remained steady though, because the disaster unfolded coincidentally as America started to increase its production of shale gas from fracking. This, in turn, freed up gas previously imported by the States.

One of the reasons the energy companies continue to explore for more fossil supplies is that, if they didn't, the price of energy would increase even faster than it is actually increasing – because of the increasing worldwide consumption and the laws of supply and demand. And the price of oil determines, to a great extent, the price of all the other varieties of energy – gas, coal, electricity etc. If the price of oil is high, then the price of all other sources of energy will be high, and vice versa – because, essentially, all the various fuels and sources of electricity all do the same thing, they provide useful energy and they are all sold on the same market.

But if there is 'too much' wind electricity being generated (for example, if the electricity demand is lower than the amount that can be generated), the price of all sources of energy (oil, gas, coal, nuclear electricity, wind electricity etc.) will drop and vice versa. At the time of writing, the cost of energy produced by large wind turbines is comparable to the cost of the energy produced by coal.

As indicated earlier in a note (and explained more fully in the next chapter), the price of a commodity is related to the labor embodied in it as well as to the laws of supply and demand. In fact it is not simply the number of hours of labor that affects the value of the commodity; it is the number of hours of *socially necessary* labor that is important.

Because it is getting harder and harder to find easily accessible oil anywhere on the surface of the planet or below it, it tends to require more and more labor power to extract it. Because oil energy is competing with wind energy, the labor time required to produce 1MWh of wind energy begins to determine the current *socially necessary* labor time required to produce 1MWh of energy wherever it comes from. The socially necessary labor time for producing 1MWh of oil increases by the day, while, because over time wind power maintenance work starts to outweigh wind power development work, and because maintaining a wind turbine will require less labor than designing, developing, building and installing it, the socially necessary labor time for producing 1MWh of wind energy decreases day by day. The price of energy begins to come down. Oil begins to become uneconomic (when competing with modern wind turbine energy). Add to that that a modern wind turbine will produce electricity with no requirement for fuel, and you can begin to see why the fossil fuel lobby is anti-wind. They will do their utmost to derail wind for as long as they can in order to preserve their vast investment in the old energy infrastructure and the vast profits which flow from it.

The energy produced by a modern wind turbine is not only becoming cheaper than the fossil alternatives; it is cleaner. A wind turbine produces clean energy, whereas the coal, gas and oil powered power stations produce dirty electricity. Each additional wind turbine connected to the grid results in a decrease to CO_2 emissions.

The financial payback time for a large wind turbine is around six years. That is, the money value spent in the manufacture and installation of the turbine is paid back on average after the turbine has been operating for six years. This assumes that there was money up front to pay for it. Of course, loans from banks to pay for the large sums of money involved usually stretch repayments at agreed interest rates over the forecast lifetime of the turbine. This could be up to 25 to 30 years depending on the technology. But, taking a simplistic financial view, after six years, the wind turbine will have 'paid for itself' and after six years, it will start producing 'free' electricity.

For conventional coal etc. power stations though, even after they have 'paid for themselves' they do not ever produce 'free' electricity because the coal, oil, gas, uranium etc., still has to be continually dug out of the ground, and the fuel still has to be paid for.

The concern for the existing traditional energy merchants, the fossil fuel and nuclear companies, then, is that the new kid on the block – renewable energy – will tend to force prices down the more there is of it on the market. This will, in turn, make it harder and harder for the fossil fuel companies to sell their commodities - oil, gas, coal etc. This is why the fossil fuel companies intervene to pressurize governments to rig the market in their favour. The 'green taxes' and 'green levies' that are being imposed on their customers by the fossil companies should really be called fossil taxes because they are levies that will enable the fossil companies to compete with the new kid on the block. There is, of course, an initial high cost for the installation of renewable energy generators like offshore wind, but, because they quickly pay off both the energy cost and the money cost (in six months and six years respectively), they rapidly become a fearsome competitor for the old industry.

Flaws in the system

Each wind turbine connected to the grid tends to bring down the price of electricity. Each wind turbine tends to reduce the profitability of the energy companies – the wind energy companies as well as the other energy companies. This is how the capitalist system works. It works in the same way in all sectors of the economy, but in general the dreary process is as follows:

The first capitalist on the market makes big profits. There is a boom. As other capitalists pile in to try to get a 'slice of the action', they inadvertently immediately begin to undermine their own efforts. Because of the supply and demand price mechanism, if an excess of energy is created, the price goes down and with it the incentive for producing it in the first place! This process goes on for years. Then, at a certain point in the cycle, there is a slump or, and, periodically, a depression. Some of the companies go bust, workers lose their jobs. The economy 'bumps along the bottom' until, stimulated by lower prices , predatory capitalists buy up capital equipment and cheap labor triggering the process to start up again – boom slump boom and on and on. These cycles vary in length but seem to last for about ten years on average. In each successive cycle, the pain gets sharper and sharper, ending in joblessness disaster for countless working families, and destruction of whole swathes of industry.

The cycle of booms, slumps and recessions sometimes ends in catastrophic failure of this or that aspect of the system. When that happens, governments intervene to bail out the stricken part, with the result that the capitalists make the rest of us pay for their crisis. That is what happened with Long Term Capital Management in the late 1990s, and more recently with RBS and insurance giant AIG.

The dynamic of capitalist accumulation and competition appears the same in all sectors of the economy, and the boom slump boom cycle described above, plays itself out across the economy as a whole including the energy industry.

Capitalists are not particularly concerned which commodity they manufacture – they are simply interested in the profit they can make. So, their investment tends to be directed at any profitable areas of the economy. This is logical for any one capitalist, but this does not produce beneficial results for the system as a whole. Because of the anarchy that reigns under capitalism, all the capitalists try to get a 'slice of the action', if they possibly can, and they all pile in. This results in over-production in that industry that previously was profitable – and which industry starts to become unprofitable as the capitalists compete and flood the market with their particular version of that product. Prices fall and the capitalists continue to fight in a race to the bottom. This is probably all good sport for some of these 'captains of industry'. If their business goes to the wall, they can simply dust themselves down, live off the stash they have in their private bank accounts, and wait for the economy to recover and simply start again with a new enterprise. But for the workers, these regular crises of the capitalist economy spell disaster. Some workers never entirely recover, others, if they are lucky, find employment again, but there is always a fragile aspect to any employment. It is always subject to the apparently random, seemingly blind destructive forces of capitalist production.

Rate of profit and rate of exploitation

Capitalism is an unstable system. This instability is built into the system. And there is a tendency for the instability to get worse over time. The slumps get deeper and the recessions sharper and more painful. This happens because there is a long-term tendency for the *rate* of profit to decline. The rate of profit is a measure of how much profit is extracted by capitalists in relation to the quantity of capital invested. In each successive cycle, because each capitalist needs to out-do the next, or suffer possible extinction, they try to produce the same or more profit as in the previous cycle, but using less workers than before with more and better (more expensive) machinery and/or systems. At each successive stage, they are forced to invest more in fixed capital. The amount of profit per worker tends to increase – workers are exploited more – the rate of exploitation goes up, but the amount of profit per unit expenditure on capital goes down.

There are two ways[15] of expressing the rate of profit; one is more useful for showing the ratio between paid and unpaid labor – more useful for showing the degree of exploitation – and is what I have called the rate of exploitation. The other is in more general use by economists. The rate of exploitation is expressed as the ratio of capital advanced as wages (also called variable capital) to the surplus value created. (I discuss exploitation more fully in the next chapter.) The second way, the one in more general use and commonly referred to as the rate of profit is the ratio of total capital, that is wages plus capital as machinery (also called constant capital), to surplus value.

The rate of profit tends to decline over time. Because of this, at each bottoming-out stage in the cycle, capitalists are more and more reluctant to invest because they calculate that they will get less and less profit from their stash of the proceeds from previous exploits. The economy 'bumps along the bottom'. Although capitalists don't necessarily plan it consciously, one way for capitalism to get out of the crisis resulting from the long-term tendency for the rate of profit to fall is for capital to be destroyed. Clearing out old capital allows room for new capital to move in at a higher rate of profit. And one way of clearing out old capital is war. Capitalism tends towards war. War is always hovering in the background; or if you are unlucky enough to live in Afghanistan, Iraq or Syria, or in many of the war-torn countries around the planet, it is right there in the foreground. This tendency towards war, for me, is the first reason why we should try to change the world. Capitalists would tell us there is no alternative to their system, but I would say there is and we must implement it in order for a safe planet to be constructed.

3) Political leadership

The report mentioned earlier, produced in 2009 for the Confederation of British Industry (CBI) and sent to the British Treasury for consideration demonstrated 'the risk that a large share of low marginal cost technologies in the electricity mix may drive low or negative electricity prices'[16]. In other words, this CBI report was admitting that if these low marginal cost technologies are employed in large amounts, the price of electricity could drop to near zero', or go negative! They were arguing the case against wind by claiming it could produce free electricity!

In the early days of nuclear power, the industry boasted that nuclear electricity would be 'too cheap to meter'. That never turned out to be correct, because of the massive cost overheads associated with nuclear electricity. But, electricity 'too cheap to meter', wouldn't be good for the traditional energy industries. Wind electricity and solar electricity (PV) are 'low marginal cost technologies'. This is because wind and PV electricity don't need any fuel. (If say demand on the grid rose by 1MW, then, if electricity from gas provided the additional power, there would be a cost – the cost of the increased amount of fuel needed to produce 1MW – whereas, if the electricity were provided by wind or solar technology, there would be no additional, 'marginal', cost – the wind is free and the sun sends no bills!). I think it is safe to assume that the CBI's motive in communicating with the British Government was not to boast about wind electricity being 'too cheap to meter' but to influence governmental energy policy and strategy in their favour. But in fact, wind energy *would* start to approach a situation where it was 'too cheap to meter' once a substantial number of turbines were commissioned.

The British government policy on energy is confused – it keeps chopping and changing. One year they are for nuclear, then they are against it, then they back it again. Policy alters with every change of administration and also as it responds to the demands of the various corporate lobbyists. The government introduces and then quickly withdraws or curtails policies (eg. the Feed in Tarriff[17]). The British government is attempting to ride two horses – that of attempting to do something about climate change at the same time as it tries to keep its business friends happy. As anybody who might have tried knows, you cannot ride two horses at the same time – especially if they are going in opposite directions! The fat horse with all the money on it, big oil and its network of friends, the banks, the auto manufacturers, the arms manufacturers, the chemical industry etc. always wins this rigged race against climate change. All governments are in the same cleft stick. That is why the Labour Prime Minister, Tony Blair, and then the Tory chancellor, George Osborne, seven years later could both come up with, practically, the very same statement about government policy on the problem of climate change.

So, Tony Blair in 2004 told Greenpeace how 'shocked' he'd been by the latest evidence on climate change, even while his ministers, with his backing, at the Department for Environment, Food and Rural Affairs confirmed that any changes (in CO_2 allowances) 'would not undermine the competitiveness of British Industry'.

And then, George Osborne, the Tory Chancellor, in 2011: 'We are not going to save the planet by putting our country out of business'.

After the Fukushima nuclear disaster in Japan, Germany

decided to not build any new nuclear plants and to wind down all existing nuclear plants. That pleased the renewable energy sector. It also pleased the most polluting industry of all – coal.

Although Germany is a country that has gone further than most along the road to clean energy, problems have apparently arisen, possibly due to the scale of the task being undertaken, with cable connections for new offshore wind turbines, and the knock on effect this is having on the installation logistics and transport. In one response to the rapid build-up of renewables installations, the Environment Minister, Peter Altmeier, recently told representatives of the German wind power industry that: 'We have too much wind in Germany' and 'Not all of your dreams are going to come true'. Germany now plans to build 25 new coal fired power stations. In a phrase reminiscent of Blair and Osborne, Altmeier explained:

'We don't want to kill the competition between such technologies [referring to the coal and gas industries], we want to encourage it so that the best most cost-effective solution prevails'. The fat horse of the fossil fuel industry wins!

And in America, as noted, President Obama did not allow the phrase 'climate change' to pass his lips once during the Presidential election TV debates with Mitt Romney from fear of losing votes by upsetting the climate change 'skeptics'. Once elected, in his second inaugural speech he made some very encouraging sounding statements about 'green energy'. But I do not think that he will be able to make any substantial in-roads against the strategic orientation of companies like Exxon and BP etc., because to do so will be to cut into their profits. For companies like these to change direction, technically, would not be a problem. (The wide skill set of the energy company workers matches the challenge of climate change perfectly.) To enable a change that would not affect their profits would. It would require a level of subsidy (buy-outs and pay-offs to the fossil and nuclear sector) that the heavily indebted US state simply cannot afford, which ordinary people would not vote for (they would be paying!), and which, bluntly, the political classes would not countenance. Capitalism is caught in a dirty, oily trap.

So the major fossil fuel companies continue to scour the earth for ever dwindling supplies of oil, gas and coal. They go to almost absurd lengths to get at the stuff: drilling miles below the ocean surface for oil; forcing chemicals down into the earth's crust in order to free gas from rocks; ripping up the earth for the dirtiest tar-rich sands of Alaska; exploring in some of the most challenging, remote and pristine areas for the last drops of black gold. All of this without much, if any, regard to the environmental and human damage it is causing.

This chapter has attempted to show where the enemy lies. The next one tries to show how it can be beaten.

Before finishing this chapter I wanted to mention EROEI. This very important concept is quite often overlooked. It stands for Energy Return Over Energy Invested. If it goes below 1, we are in trouble.

I've also very briefly mentioned the Agriculture and Construction sectors, since, as well as burning vast amounts of fossil fuels these industries contribute to global warming through different mechanisms.

EROEI

The financial payback time for a large offshore wind turbine was briefly touched on above. Incidentally, and importantly, the *energy* payback time for a large offshore wind turbine is considerably less than six years – it takes between three and twelve months for the wind turbine to generate the equivalent amount of energy to that used up in its manufacture and installation. The concept of energy payback or its more formal measure – EROEI[18] is extremely important for the future of energy production. Essentially, the EROEI for wind and PV is much greater than the current EROEI for fossil fuel. EROEI is a measure of the energy intensity of energy generation, a measure of how much energy is used in the production of energy itself. EROEI must not drop below 1. An EROEI value lower than 1 would mean that more energy is consumed in the energy generation process than is generated. The EROEI for fossil fuel has been steadily declining over the years because the remaining reserves of oil and coal and gas require increasingly more energy to be expended in their extraction. If nothing is done to stop this decline in the EROEI (for example by switching rapidly to wind power, which has a relatively high EROEI) there will possibly be major problems for the capitalist economy as the percentage of GDP going to energy production increases at an accelerating pace[19].

More greenhouse gas

The main greenhouse gas is carbon dioxide and the main source of it is the burning of fossil fuels. That is the reason why I have concentrated, in the technical bits of this book, on showing how CO_2 emissions from burning fossil fuel can be eliminated. The burning of fossil fuel accounts for more than 90% of the anthropogenic (caused by humans) CO_2 emissions and eliminating the burning of it will be a huge step in tackling the threat of climate disruption. CO_2 is not the only greenhouse gas though, and burning fossil fuel is not the only source of atmospheric emissions.

Earth

More than 5% of the CO_2 pushed into the air as a result of human activity gets there as a result of modern cement production. In the cement making process, chalk and/or limestone rocks which have remained inert for millennia are heated to very high temperatures. This process creates the ingredients for cement making, but it also produces CO_2. The CO_2 is driven off in an irreversible chemical reaction.

The main use for cement is for producing concrete. Concrete is an immensely strong and versatile material, and it has allowed for the construction of structures like bridges and large buildings etc which without it would have been impossible. Concrete use could be kept to where it is an irreplaceable material – to where using its strength and flexibility is essential. But the building industry often uses concrete where another less polluting material could be used in its place. Cement manufacture, like the oil industry, is an industry central to capitalism. It will be difficult to deflect it from its intransigent 'business as usual' stance. Again, as with oil and with any other industry, profit comes before planet, and therefore no systematic attempt is being made to change building practice.

The Romans, over two thousand years ago, discovered how to produce concrete which set hard even under water. The basic chemical process that went into making the cement, however, was different to that used in most (Portland) cement now. The difference mainly was in the much lower temperatures they used in the process. That different process resulted in concrete that cured over a longer period of time. That curing process involves the re-absorption of CO_2, thus making the whole process carbon neutral. For most building work, ie where the superior strength of modern concrete is not required, using a traditional low temperature lime based cement would do just as good a job, and would be kinder on the environment. The magnates of the building industry though have hit on what they see as the most profitable way of throwing up buildings and houses etc and they plan to stick to that method, it seems to me. But there are plenty of other building methods that are simply not considered or countenanced by the magnates. Rammed earth, for instance, is nowhere near as strong as modern concrete. But you do not need the strength of modern concrete to put up buildings up to 3 or 4 storeys high.

Rammed earth requires careful selection of subsoil. The

subsoil is placed in a suitable frame and is then 'rammed' either by hand or machine to produce rock hard walls.

Very little, or no CO_2 is produced by the rammed earth process. Rammed earth buildings are very rare at the moment, but rammed earth was one of the most common ways of building small structures all over the world up until relatively recently. Thanks to the perseverance of 'pioneers' like Rowland Keable[20] rammed earth has recently been recognized and, for instance, is now included as an authorized building material in the UK Building Regulations. A seven metre high rammed earth wall is the centre piece of the award winning WISE building in Wales[21].

Building with other more traditional materials, such as straw bale or wood, can, done with care, actually sequester carbon. Expanding the repertoire of building methodologies in use would also provide many more jobs.

Agriculture

It is argued by the defenders of capitalist intensive farming that there is no alternative to intensive methods and practices such as GM, but there is a small army of agricultural experts who have produced evidence to the contrary[22].

Politics

As with fossil fuels though, the existing construction giants and food producing multi nationals will not tolerate any change if it threatens profits.

Again, the problem is not technical know how, the problem is: How can these technical solutions actually be implemented. The problem is political.

Chapter 5

Identifying the Agent for change

Rise like Lions after slumber
In unvanquishable number,
Shake your chains to earth like dew
Which in sleep had fallen on you -
Ye are many - they are few.
Shelley - The Mask of Anarchy, 1819

I was standing at a Waterloo Station underground entrance watching the workers stream in. They were heading for their various workplaces in the city – to nurseries, offices, hospitals, shops, schools and factories etc. – a continuous torrent of thousands of women and men every minute, from early to mid morning. Of course, there is a similar daily early morning scene of people hurrying in to work enacted everywhere – in all of the villages, towns and cities worldwide. And, in my mind's eye, I saw them rushing in, in their billions, to workplaces in farms, towns and cities across the UK and across the planet.

I was looking at where the money comes from.

Work creates wealth.

Those workers that I was watching create all the wealth.

I will try to show how this works below, but why is the question of how all the wealth is created important, in a book about how to make the world a safer place?

It is important because an understanding of this, and concomitantly, of how the capitalists increase their wealth, helps to identify the internal structures of the system which the enemy within, identified in the last chapter, will be determined to defend. It exposes the deceitful, robbing, reckless, bloody heart of the beast which is destroying the planet and making a misery of the lives of billions of us human beings. I try to show in this chapter, by looking at a number of subjects that might seem unrelated, some of the dynamic of capitalism and, importantly, the forces and methods on our side that can make it possible to slay the beast.

The capitalist economists used to recognise that wealth is created by labor, and perhaps, some of them still recognise that this is and always was true. But, actually, the capitalist economists have shown, clearly, that they do not understand how the capitalist economy works. If they did, they might have known, in early September 2008, before the collapse of Lehman Brothers Bank, and the subsequent bail out by governments of other failing banks and insurance companies, that their capitalist 'market' had failed again (or rather it was doing what it always does) and that another recession was heading towards them like an out of control driverless freight train. What is doubtless also true, is that even if they had seen it coming, they and their class, would not have known how to stop the inevitable crash and the ensuing misery and joblessness that it would bring about. The capitalists and their economists do not understand the capitalist economy.

Their capitalist beast is, and always was, out of control.

Can this beast be tamed?

I think the record shows that capitalism cannot be tamed or reformed, and so, if it cannot be tamed, then, because of the increasing misery and ecological damage that it causes, we have an obligation to change it. We must get rid of the capitalist system and replace it with a system that allows all people to live full, joyous and fruitful lives, with a system safe for people and planet.

So this chapter is about how this can be done. It is about identifying the agent for change.

Capitalism is forcing a confrontation. Since its inception, the capitalist system has been, through forces entirely consistent with its internal laws, creating the conditions and building the agent that is historically destined to be its destroyer.

This agent is the class of people brought together and exploited by profiteering capitalists.

Capitalism came into being in a revolutionary process.

It ushered in a completely revolutionary form of production. This system of production was very different and much more dynamic than all of the preceding production systems – primitive communism, slavery, medieval production.

The new production system was social in that it brought people (soon to be called workers) together to produce. But capitalism didn't revolutionise everything. It left some of the old systems intact – principally, the system of exchange. The new capitalists appropriated product for themselves and sold it using the old systems of exchange just as they found them.

Those systems have been in existence now for about 400 years. They have passed their use by date. The economic and political and ecological crises we can all see before us are indicative of an unstable system, unable to deliver a safe path forward. All that is on offer is the prospect of unemployment, debt, conflict and war. People experience these aspects of capitalism every day. But amongst this, hidden, almost sleeping, is the force which can burst asunder the chains binding us all to that instability, a force which is held back by the destructive laws of capitalism. That force is the active organised working class.

Capitalism will not just fall over and die all by itself. Like an old oak tree, it will stand there forever unless it is chopped down.

The best analysis of capitalism and explanation of the historic role of the working class is, as I repeat below, to be found in Marx and Engels' work – principally, 'The Communist Manifesto', 'Socialism, Utopian and Scientific' and 'Capital'.

What I am attempting here is to wield the tools of Marxism, to paraphrase, as best I can, what I have come to understand are guides to action from the best guides we have.

So several of the following paragraphs talk about what I see as an important idea which flows from Marxism: Exploitation - which, together with an associated concept - the tendency of the rate of profit to fall - is the cause of the repeated boom slump cycles of capitalism. I attempt to relate this to what real workers are experiencing now, and I go on to state what some might see as the obvious, but which I cannot see any cleverer way of saying. Organize in your workplace, agitate in your community, form or join a union and form or join a workers' party.

Know your enemy

So, to be able to chop down capitalism, I think it is necessary first to understand where its strength lies, what are the forces that govern its motion, its internal dynamic. An understanding of the life force of capitalism will enable an understanding of how and where to strike the fatal blow. 'Where the chains of capitalism are forged, there must the chains be broken'[1]. Very fortunately for us, the vital analytical work which can lead to an understanding of the life forces of capitalism has already been done.

The labor theory of value was first set out centuries ago by two capitalist economists – David Ricardo and Adam Smith. Smith and Ricardo had a much clearer understanding of how the system works it seems, than our more modern apologists for capitalism – the economists at the London School of Business, for instance[2], who happily advise various governments about money and economics.

To repeat, far and away, the best critical analysis of capitalism was done by Karl Marx and Frederic Engels. They took the labor theory of value of Ricardo and Smith and combined it with other socialist theory, developed it, and produced work which is invaluable to anyone interested in obtaining a clear understanding of the inner workings of the capitalist system. They didn't want simply to understand capitalism though. They wanted to get rid of it[3].

Money

Understanding where the money comes from is crucial in both understanding how the system works and also in generating a desire and the means to change it.

Money is the ultimate or universal commodity. It can be exchanged for any other commodity. Its value is that it can be exchanged for any other commodity. It can be said that it is the repository of exchange value.

Commodities have two kinds of value. They have the value that I am mainly going to talk about here – exchange value. But they also have another, more obvious value – their use-value. They must have use-value or they can have no exchange value. Use-value simply means the thing will have some use, it can be used; eg, the use-value of a chair is that it can be sat on. And, although I said earlier that work creates all wealth, it does not create all value. The natural world we humans are born into, of course, provides value. The wealth I spoke of at the top of the chapter is exchange value.

Under the capitalist system, everything becomes a commodity that can be bought and sold – everything can be exchanged at a price or for a certain amount of any other commodity. Incidentally, the reason that very different commodities can be exchanged for one another is that they have something in common – that something is the labor expended in their manufacture. Everything has an exchange value and that value is related to the quantity of labor required to manufacture it. Workers sell their labor power to an employer in exchange for a wage. Labour-power is also a commodity on the market that is exchanged for money. The price (the amount of exchange value that something exchanges for) of every commodity is determined on the market by the laws of supply and demand. And the price of the commodity labor power (wages) is also determined in this way. The wage, the price of labor, is determined, *before any work is done,* through negotiations between the worker and their capitalist employer. A unique property of the commodity labor power is that it produces more value than it costs to buy. This is true. If it were not, capitalists would not purchase it.

But how does this work? How does labor power create more value than it costs on the open market? The answer is suggested by the question.

Wages are determined on the open market – **in the realm of exchange**.

The value that workers create is produced *once they start work* – in a totally separate realm – **in the realm of production**.

The capitalist faces the worker with a grin in the realm of exchange and with a scowl in the realm of production. Anybody who has been a victim of disciplinary action at work or has been sacked or has been on strike will know the scowl. The grin appears as they fleece you when you try to buy what you need on the open market.

Divisions of labor

When people worked together in many pre-capitalist societies it was on a spontaneous, unplanned basis. However, in any capitalist workplace or enterprise, there is a *systematic and highly organised* division of labor. The work is organised so that the optimum or maximum productivity or production output (whatever that output might be – toddlers looked after, buses running on time, services delivered, widgets manufactured, contracts completed, lattes prepared etc.) is delivered or achieved.

The price that this product is sold for bears no relation at all to the wages paid to the worker. The wage was negotiated earlier.

Many hands make light work; two heads are better than one. Since the dawn of (human) time, people learned this and worked together and/or co-operated to produce. In the first (primitive communist) societies, people worked together socially to produce what they needed and the product (collected wood, carried water, boiled water, picked berries, killed or captured prey etc.) was shared out socially. People working together are a lot more productive than one person working alone. Capitalism uses this quality and brings people together socially to produce. But the capitalist, not the associated producers, appropriates the product privately. Wages are handed out according to the previously agreed rate. The difference in value between what the capitalist appropriates and what he pays the workers at the agreed rate is value that the worker has created over and above the cost of labor. This is where the wealth comes from.

Surplus value

If the company is profitable, the exchange value the capitalist appropriates will be greater than the total wage paid to the workforce. This surplus value (the exchange value remaining after all the capitalist's expenses such as wages and rent etc. have been paid) is pocketed by the capitalist and becomes the capitalist's profit.

Labor power has created more value than it cost on the open market.

If the company is not profitable, it will eventually fold with the result that the workforce will lose their jobs and they will add to the reserve army of unemployed workers.

So, in a 'business as usual' situation where the company is making a profit, workers do not ever get paid for all the value they create. If they work for eight hours, they might receive value equivalent to seven hours', or six hours', four hours' or one hour's work, depending on the wage rate negotiated at the start of their employment or, putting it another way, depending on the rate of exploitation. The wage paid bears no relation to the value obtained by the capitalist for the commodity produced by the worker.

Workers are always exploited. *They don't get paid for all the work they do. No worker does.* The capitalist sells the product, he (and the majority, though of course not all, of the big capitalists are men) appropriates from the workers, on the open market for the maximum possible price. The work done by the workers has generated enough value to cover their wages, the capitalist's taxes, the capitalist's rent *and* the capitalist's profit.

Crisis

Worker exploitation is the source and engine of capitalist growth. Going back to that rush of bodies at the beginning of this chapter; every now and then a capitalist slips by (although they are more likely to be found in their limousines or helicopters and private jets). All the money and wealth produced by the rest of us goes through their bank accounts. It is as if the workers are pumping value out of themselves – value which is seized from them by the capitalist the instant it is created. Of course, the commodities produced by their work, must be paid for in the main by other similarly exploited workers who exchange some of their wages for the commodities. The total amount of wages can never match the total value of the commodities they have produced. This applies whether the commodities produced are tools and machinery for other capitalists, luxury goods for the rich or whether they are commodities destined to be bought by other workers and this is the root of one of the instabilities in the system and which eventually leads to crisis.

Exploitation leads to needs not being met at the level of the worker; and also to the instability of the entire economy at the opposite end of the scale. The capitalists scour the world looking for buyers for their products. But the purchasers they are seeking out cannot buy all these products because they are other workers who have been robbed by other capitalists of the total product of their labor. Over time there comes about what is called, a crisis of overproduction. The capitalists are unable to sell all of the commodities produced and the system slides into a slump.

In a time of slump, the surplus value produced in the earlier phase, the profit, is hoarded by the capitalists, by the rich, by capitalist owners who have 'lost confidence' in their ability to produce and sell whatever commodity it was they were involved with. In other words, because of the tendency towards 'overproduction' (it is called *over*production, even though there might well be great need for that commodity), they say there is 'a lack of demand'. They are unable to exploit any more human labor at that time. Factories close, workers are laid off. The money sits in the bank accounts (and other asset classes) of the rich. The workers have produced 'too much' and because of this, they lose their jobs or suffer a cut in wages. As Engels put it 'the working masses lack the means of subsistence because they have produced too much of them'[4].

So, exploitation not only robs the individual worker of the fruits of her/his labor; it also is the cause of the massive instabilities in the capitalist system – the tendency for booms and slumps in the economy. The system attempts to expand – in the boom phase, but, sooner or later, the fuel of demand runs out because of low pay[5] resulting from the exploitation of the buyers (the buyers, in the main, being other workers employed by other capitalists) – and then it enters a period of slump.

After a period of stagnation, after a time in the slump phase, after mountains of unsold commodities and productive tools have lost a substantial part of their value or been destroyed, after factories, offices, and retail chains have closed, some of those capitalists sitting on hordes of cash accumulated in the preceding phase, reckon they can start to exploit workers again and bring capital and workers together yet again to begin yet another phase of the capitalist cycle.

Fight

Of course, workers are not purely victims of this process. They fight back. Capitalism brings them together in workplaces. They get to know each other, loyalties are built up. If and when it comes to a fight with the boss, an organised workforce can be the most powerful force in the world. When organised, they can force the boss to pay them more, and they can demand better working conditions. This is because the capitalists need the workers. Because of this, because the capitalists need the workers, the capitalists can be forced to concede. What the bosses know, and what the workers have not yet learned though is that **the workers do not need the bosses**. Workers can learn this in the struggle.

As long as the system remains there will be a battle between bosses and workers, with the boss class trying to maximise profits and the workers attempting to win better wages and conditions – the class struggle.

Competition between capitalists attempting to maximise their profit means they try to keep wages as low as possible, whilst, at the same time they try to sell their commodities at the maximum price. Workers are hit both in the realm of production and also in the realm of exchange[6]. Competition means workers are paid less and less while the essentials they need to buy to live on become more and more out of their reach. Their 'own' capitalist keeps their wages low, the rest keep prices high.

Limits

The commodity labor power is unique in that it is the only commodity that can generate more value than it cost to buy. There are limits to the amount of surplus value that can be generated, however. The labor theory of value developed by Engels and Marx shows that the exchange-value of a commodity is proportional to the amount of socially necessary[7] labor time required to manufacture it or that is embodied in it. For the commodity labor power then, this exchange-value must be derived from the socially necessary labor time required to 'produce' the worker – the sum of the daily subsistence costs for the worker: food, shelter, clothes etc. plus the costs of 'producing' the next generation of workers (workers bringing up children).

So, capitalism cannot reduce wages beyond those basic levels of subsistence without endangering the ground on which the whole edifice stands. Capitalism needs the existence of an army of healthy and educated workers. But individual capitalists ***do*** **attempt to reduce wages beyond this** in order to maximise profit. The tendency is therefore for wages to become ever lower and lower. And for capitalism to begin to undermine the ground it stands on.

In countries like the UK, it is clear that some wages have reached rock bottom (indeed have gone below the basic subsistence levels) – since the state steps in to support a significant number of full-time workers whose pay is not enough to pay for basic subsistence – by paying for an element of housing and subsistence costs (Housing benefit and Working tax credit). This state support does not provide security for these workers however, since the eligibility for these 'benefits' is constantly being eroded for all but the most severely disabled workers. And even the benefits of severely disabled workers are constantly being cut back, thus depriving them of the bare minimum for civilised survival.

Exploitation

The crux of the problem is that, under the capitalist system, commodities are produced socially but appropriated individually (privately) by capitalists. It was only when I understood exploitation, as uncovered by Marx, that I formally became a socialist. Up until then I had had the idea that there was something slightly sleazy, slightly distasteful about any thoughts about 'starting a business' and getting people to work for you. I didn't know why I didn't like the idea – I just didn't. Discovering exploitation provided the explanation of that unease. Getting people to work for you meant you were going to live off their blood sweat and tears. To make a profit, there was a good chance that you might create misery in their lives as you forced down wages in order to maximise your profit.

Capitalists, apparently, don't suffer from that sense of unease.

Initial capital

Of course, the capitalist brings an essential element. He provides the initial capital – the buildings, tools and wages which allows the process to begin. The apologists for capitalism argue that any profit due is coming because of this risk they take with their initial outlay.

However, once the work has begun, the profit accumulated from exploitation quickly balances the initial investment amount. From then on it is pure profit. As discussed above, exploitation produces enough value to cover rents, costs of training and tooling, payments to other capitalists, taxes, wages and the initial outlay; in other words, all of the expenses of the capitalist, and it produces more and more profit until the business once more becomes a victim of the random anarchy of capitalist production when, again, competition from rival capitalists restricts the flow of profit.

At this point as mentioned above, either workers are sacked and the business folded and sold, or the business attempts to brave the storm by battening down the hatches by attempting to put the workforce on reduced wages and short-time working, in a last ditch attempt to conserve capital until the upturn arrives. And in any case, in the majority of situations the initial capital outlay comes from profit accumulated from a previous cycle of exploitation, and which previous cycle was entered into precisely to accumulate capital.

So, is there an alternative to this madness? How can the outrages at the clothes factory in Dhaka, Bangladesh or the killings at Bhopal chemical factory in India or the blundering waste of life and ruination of the seas on the Gulf of Mexico or the countless other crimes carried out by capitalists at their workplaces be prevented? And what has any of this got to do with renewable energy and climate change?

In short, the illusion of democracy that many of us live under must be replaced with a real full democracy. When we have control over what we produce at our workplaces, then we will be able to control the safety at our workplaces and in our environment. But how can we achieve real democracy? Obviously there is no easy answer to that - otherwise we would have done it by now! First, we must try, and we must try to learn from other workers struggles and from our history of struggle.

The death factory workers were forbidden to join a union. There are many reasons why capitalists detest unions and try their utmost to keep work places union-free. For one thing, workers' wages in unionised workplaces tend to be higher (by around 20%[10]) than wages in non-union enterprises. Unionised workplaces are much safer than non-union workplaces[11], and safety will cut into the capitalist's profit. Bargaining collectively for a wage increase is a much more effective negotiating strategy for the workers than a situation where workers face the boss individually and where the capitalist, using divide and rule tactics, can set one worker against another. The bosses attempt to prohibit collective bargaining. They try to create a weak and divided workforce by, for example, paying men more than women, by paying workers in a plant in one area at a better rate than those in another plant elsewhere etc, by paying office workers more than physical laborers. Unions have the potential to obstruct these divide and rule tactics.

Ideas which allow the bosses to divide us must be challenged – sexism, racism, homophobia, are all tools of the boss class. They use them constantly, but workers' solidarity and unity on our side can defeat the cruellest boss.

The vast majority of people have no choice but to find work with a capitalist in order to get enough money to live on. Of course, the primary reason, under the capitalist system, for a worker to find work is for them to acquire the means for their subsistence – to get money to pay their rent or mortgage, to buy food, pay the bills etc. It would make no sense for them to choose to work in a dangerous place, because if they were injured at work, or worse, lost their life, this would not only be an immediate tragedy for them and their dependents, it would, of course mean that they lost their means of subsistence. No worker wants to work in a dangerous workplace. But, as we know from tragedies like the one in the Bangladesh clothes factory, and many others[12], millions of people are, every day, **forced** to work in the most appalling and dangerous conditions in non-unionised workplaces in order to simply get enough money to prevent them and their families from starving. The workers at the Rana Plaza clothes factory and many like it were and are not paid good wages. Their wages do not even approach a living wage. The slums they are forced to live in, huddled up next to the factory (just like they were at Bhopal in fact) would make you cry with anger and many other emotions. Why? Because of the appalling nature of those conditions combined with the fact that they are simply unnecessary.

A rational, democratically planned economy would be able to provide a safe working environment, good living conditions and a decent living wage for everyone. Capitalism will not provide this, because the driving force for capitalists is production for the maximisation of profit, not production to satisfy need.

An organised workforce can force the employer to provide the necessary safe conditions for work; they can strike for better wages. Individual workers cannot go on strike – they will simply be sacked. Capitalists know that an organised workforce can simply refuse to work in dangerous conditions. But the problem for the existing individual workers in a non-unionised workplace is that they know they will be sacked and will lose their livelihood or worse for simply talking about unions or organising. Starting out to organise and unionise a non-unionised workforce is a very difficult (and in some parts of the world, very dangerous) thing to do.

But this highlights what I called the crux of the matter earlier. The capitalists do not want you to do it because they see their profits threatened, and they will rigorously oppose any attempts at unionisation. It comes back to this: under the current economic system, there is *social* production of commodities, but the product is appropriated *privately*. Any incursion into the private appropriation will be resisted with the utmost force.

The exploitation going on in the developing countries like Bangladesh and India is a kind of super exploitation. The differing exchange rates between the country where the product is manufactured and the country where it is finally sold mean that the capitalists can rake in super-profits. The profits are magnified by the exchange rate differences. The global nature of commodity production now means that the final product might be comprised of elements manufactured in many parts of the globe. Profits are globalised and misery regionalised. The same capitalist can be exploiting people in many different countries at many different rates. But there is no fundamental reason why things could not be organised very differently to this. In a nutshell, what needs to be achieved is this: Production must continue to be social, but private appropriation must be replaced with social appropriation.

That would of course be a social revolution.

Change on this scale would not happen overnight. Revolution is a process not an event. The revolution in Egypt is still ongoing more than two years after it began – the result is in the balance and has not been written yet. What I can say about that revolution is that it was not until the workers organised and went on strike and marched, that Mubarak was toppled[13]. This is the main lesson: Workers hold the key to the future society. Only the working class can break the knot with which capitalism holds us all bound. We must fight, not only in what could be called the realm of exchange - workers and unemployed workers and communities and other working class organisations combining outside the workplace – with eg, demonstrations, rallies, meetings and marches. We must also fight in the realm of production – with strikes, occupations, picket lines, working to rule etc and it is important to recognise that the core power of the working class is *in* the workplace. Ultimately, the working class must seize control of the means of production, and start to produce for need. When the workers try to do this, as they would be doing in first small steps by forcing capitalists to provide safe a working environment in the clothes factories of Dhaka for instance, they would find that the forces of the state – the police mainly, and then possibly the military, will be used to oppose any attempt at change. This has been seen over and over again whenever the rule of capital is challenged. The police always support the bosses, never the workers. In a strike they will be seen always trying to defeat the working class. So workers cannot simply take over the means of production (hospitals, schools, factories, burger shops, coffee shops, supermarkets, bakeries, banks, oil refineries, shipyards etc.) although this will have to be done; in our struggles we will have to challenge the forces of the state to win our demands, and, in the final struggles if we are to rid ourselves of capitalism once and for all, we will have to also defeat the existing state which protects the existing order.

Freed from the chains of capitalism which ensures that subsistence provision is held up because there has been too much production, a democratically planned economy can be developed based on the needs of everyone. The private accumulation of profit for accumulation's sake by the capitalists will have come to an end. Freed, people would be able to develop rational solutions for housing, health, energy production and agriculture etc. – things which now only get developed if profits can be made. The partial democracy of today will have become full democracy – where all the people's needs determine the outcome, the outcome previously decided by the greed of a few.

Both kinds of humiliation – extreme poverty and deprivation, and the degradation of the planet's ecosystem – have the same cause and they both have the same solution.

The cause is capitalism.

The solution is for the exploited class to break the back of the capitalist class, to take political control and to seize control of the means of production – nurseries, schools, hospitals, factories, farms, offices, etc., etc. – in order to run and to build a planned economy based on the needs of all of humanity.

As discussed earlier, the capitalists need us. We don't need them!

Class

For any of this to happen, the exploited class – the working class by another name, has to recognise the nature of the exploitation, become conscious of the class nature of the exploitation and start to operate as a class.The good thing is, we, the working class, are the vast majority – the 99%.

A lot of workers think they are not working class. They might work in a nice office or salon for example and think of themselves as 'middle class'. All sorts of things are said about class. Recently there was a study in the UK which maintained there were now seven Classes! Other people say that your class is determined by your accent or what kind of school you went to.

No.

These kinds of misleading ideas have existed since the inception of capitalism. It is the interests of capital to sow doubt and confusion over the issue. I don't think many people study revolutionary socialist ideas at school. There is a reason for that. People might start to think they can change the way things are. They might start to think for themselves.

If you have to go to work in order to get money to live – if you sell your labor – then you are a worker and you are a member of the working class. If you employ people to make a profit from them, you are a capitalist.

There are very small grey areas, border lines, of course, but the two main classes are us, the workers, and the exploiters, the bosses. For instance, perhaps because they cannot find a job, some people become small capitalists (sometimes referred to as the petit bourgeoisie, as opposed to the real capitalists, the bourgeoisie) – small shopkeepers etc. These people want to become big capitalists, but they cannot get there because of the nature of the competition. Those small capitalists can still feel strong links to the working class and their destiny is tied up with the working class. They could be classed as fellow travellers with the working class. They might find themselves playing on a 'sticky wicket' – because, to mix my cricketing and equestrian metaphors – like a circus act they are riding two horses at the same time – a difficult trick to accomplish. The people in the 'petit bourgeois' class – ie. those who are neither capitalists nor workers – who are neither employed, nor do they employ anyone, can be described as the middle class. The dynamic of capitalism also drives a lot of these petit bourgeoisie back into the working class.

Many who are in the middle class – musicians, writers, small shop keepers etc., feel a strong attachment to the working class, and support working class causes and movements and, rightly, consider themselves part of the working class movement.

Capitalism ensures that there is, as Engels described it, an industrial reserve army of unemployed people – it helps the capitalists to pay low wages to those lucky enough to find work; those tossed on the scrap heap by capital, the unemployed, are an integral part of the working class. The industrial reserve army created by the deranged system upheld by the capitalists can, though, be a crucial force within the working class, helping deliver ultimate victory. When, some years ago, I was on a picket line outside the Girobank, I didn't really understand why some socialists had turned up selling a newspaper. There was no automatic mental connection between fighting the boss (Girobank) and fighting the capitalist system. It was only some time later that I realised that the socialists selling that paper and us workers on that picket line were fighting the same fight. This consciousness applies in many situations. A lot of unemployed youth (and the system has created a lot of them, especially black unemployed workers - who the system systematically discriminates against) often do not regard socialists as having anything at all to do with their struggles against society. Nothing could be further from the truth.

Internationalism

Capitalism is a global system. Any attempt to conquer it must also ultimately have a global reach. It would not be possible to, say, rid one country of capitalism. This was tried in Russia at the beginning of the twentieth century. The rest of the capitalist world turned on the incipient socialist state and crushed it. Russia was forced to compete (especially militarily) with the west (the rest of the developed world at the time). The attempt to out-compete the capitalist west by the very much weakened post-civil war Russian state could never result in socialism or communism – the result of that military competition between states was state capitalism in Russia[14]. Before the end of the 1920s, workers in Russia were exploited too. That exploitation took on a different form. The bureaucratic central command economy set up to compete with the west killed the shoots of the real workers' democracy that was possible in the very early years of the new post-revolutionary Russia. Up until now, we have never seen socialism or communism anywhere on the planet.

Opponents of the left often attempt to characterise 'the left' as knowing only what we are against, not what we are for. This, of course, is simply a lie. I consider myself to be on the left; to be a socialist, indeed, to be a revolutionary socialist. I want to see a world free of poverty, free of war, where the resources created by people are used to build a safe, healthy environment for all; a world in which the people's artistic and scientific creativity is set free. Free from the grinding necessities of fighting simply for survival which characterises life for the vast majority of humanity under the yoke of capital.

Some opponents (almost always the very rich) also say that we are jealous of their wealth. That is not at all true, and is of course an insult. I am not jealous; I am angry. I am angry at the exploitation. I am angry about the destruction of the environment. The super-rich are super rich only because of exploitation. They are rich because millions of others are made poor by the systems the rich uphold and because they are able to destroy our environment seemingly with impunity..

Deficit

Currently, in this seeming never-ending crisis of capitalism, in the UK, and also in many other countries, we are living under 'austerity'. We are told things like 'we are all in this together' everyone must be squeezed financially so that 'Britain', 'the country' can recover, get out of the recession, and return to 'growth'. There is a 'deficit' we are told. By this, what the politicians and their tame economists mean is that there is a gap – between what 'we' (the country) spend, on everything – health, bombs etc. – and the revenue the government collects.

Well, I say, we, and by *we* in this case, I mean the 99%, the working class – we need to reflect on what they are saying regarding a deficit. There *is* a deficit. But it is not the deficit the government talks about. It is the difference, the gap, between what we create and what we receive in return for our work. Some of what we create goes to our wage packet. Some of it goes to government taxes. The rest of it goes to profit. And, that profit is not under our democratic control, it is used by the capitalists to do what they want. Frequently, it is used to exploit another set of workers. Sometimes, in times of recession, it is simply hoarded by the rich.

The deficit governments talk about could be cleared overnight if they took the wealth of the super-rich. But the ruling class in all countries is trying to clear its accounting deficit by making *us* pay. Governments worldwide are doing all they can to allow the rich to get away scot-free. So, in the UK, while cutting benefits for the poor, the government cuts taxes for the rich, it caps benefits, but shies away from capping the bankers' bonuses. It allows some companies to pay almost no tax at all and carries on handing our money to arms companies that deliver totally wasteful fiascos like the trident nuclear bomb programme or spying drones that direct long distance murder.

The fight for a decent living wage for all is very important as it is a popular call and will challenge the capitalists in their pockets. It is not a coincidence that, in the UK, the number of 'zero hour contracts'[15] went up by a quarter last year, while at the same time millions of families had to resort, at the end of each month, to 'pay day loans' and high interest loans like overdrafts (actually they should be called *No pay day* loans) *simply in order to buy food.*

Zero hours contracts seem to be getting popular with some companies. They claim that these contracts offer flexibility to workers. But what they are, it seems to me, is a return to the kind of workplace practices that workers have fought against and driven out decades ago. Dock workers used to have to turn up at the docks not knowing if they would get employed that day. If they were lucky, the foreman would select them, otherwise they would have to go away with no work – and no money.

Zero hours contracts are the same. You do not know if you will be called or not called by phone by the employer to work for, say, 9 hours the following day. If they don't call you, you don't work and you don't get paid, if they do call you have to turn up for work. That means you cannot plan anything else for the next day. But these are not second jobs the person does for a little extra pocket money. They are the main source of subsistence.

So if they don't get work, in order to eat, some people will have to go to the loan shark capitalist and borrow money from them and take all the misery which goes with that.

Its almost as if the capitalists have decided that, since workers wages are so low that it will be impossible for them to buy all of the product produced by the working class, then they will force workers into debt (to yet another capitalist), lend them money so that all the commodities can be exchanged for money. Because the capitalists are forced to squeeze more and more surplus value from the working class, or face extinction, the system creates misery and it runs on debt. And the growing indebtedness will inevitably contribute towards another and more extreme crisis.

First things to do:

Join the existing union at your workplace. If there is no union, with great care and secrecy, very cautiously and patiently, try to get your workplace organised and form a union. Join a workers party.

The party

The unions and their leadership do not want to see capitalism destroyed. They are not revolutionary organisations. Trades unions work well as 'centres of resistance'[16] against the encroachments of capital. Unions will fight for better terms and conditions at workplaces - that is their job and it is why every worker should be a member of their workplace union, but they will baulk at going further.

To go beyond capitalism, the working class needs to be united and focussed, it needs organisation over and above the unions, it needs a party.

Big business, capitalism, has its political parties – the Republican party and the Democratic party in the US, the

Conservative party and the Labour and Liberal parties in the UK. Other countries have similar organisations.

What about the workers' party? The question of what kind of party we need, or whether we need a party at all is the great question of our time. The working class has the potential to forge a better world for everyone: after all, we already make everything; everything manufactured has been created through the mental and physical labor of workers. Whether workers realise it or not, they are part of a class of billions – the working class makes up the overwhelming majority of humans living on the planet today. The most important question as I see it is: what can we do that will enable the class to *act as a class?*

Why have I put the British Labour party in the capitalist camp? Let me quickly explain, before I am in danger of throwing the baby out with the bathwater.

The unions in Britain created the Labour party in the early part of the twentieth century (the great socialist, Tony Cliff, founder of the International Socialists, characterised this very amusingly and correctly: 'Labour came out of the bowels of the unions'[17]). It was an attempt by the unions at building political representation for the working class. Although Labour (or New Labour as it has called itself) claims to represent workers, it acts like a party of big business in many ways. The working class have voted for Labour in their millions and joined it in their tens of thousands. Right from the outset though, the Labour party was never fully behind the workers. It always compromised and capitulated to capital. You don't really need to look any further than the Labour Government of Tony Blair, and the Democratic Party under Clinton, and although he is a vast improvement on the presidents that came before, President Obama's Democrats, to see that these parties are not parties of the working class. Although millions of workers look to Labour, and it is a fact that some great reforms were won (like the National Health Service), Labour is not and never has been a workers' party; it is a capitalist workers' party. The working class needs to form its own party so as to provide an organisation that can help it act as a class for itself. We need a workers' party, a revolutionary Marxist party, an Engelsist party – as Engels phrased it, a scientific socialist workers' party.

We need to build political parties everywhere and in every country which are led by workers and which work for the working class and which will fight for a better world for all. I am not going to be prescriptive about what particular party must be built in each country and how it can be built – that is up to workers wherever they may be, and is not the job of this book, but one thing is clear, on our own we are weak, together we can win. And remember these simple words, paraphrasing Trotsky:

With the workers? - Always!

With the vacillating union leaders? - Sometimes!

With the bosses - Never!

Chapter 6

Enter the Dragon

Writing about energy without mentioning nuclear power is like ignoring the proverbial elephant in the room. Or, more poetically perhaps, ignoring nuclear is like ignoring a foul smelling, fire breathing dragon in the room.

Nuclear

Why not use nuclear to produce all our energy requirements? I think there are at least seven main reasons:

It's not safe,

It's not cheap,

It's not necessary,

It's not reliable,

It's dirty.

It provides material and expertise for the nuclear bomb industry.

There are two types of nuclear – nuclear fusion and nuclear fission. Both are problematical. A nuclear fusion reactor has been in the planning stage for decades. Although billions of euros, dollars and pounds have been spent on developing this technology, power from human nuclear fusion is still a pipe dream. The technology is nowhere near the 'commercial' stage.

UKAEA

Nuclear fission is what most people mean when they refer to nuclear power generation. Most fission reactors consume the fuel Uranium 235 (^{235}U). To add to the above litany of reasons to be worried about nuclear technology, the world reserve of ^{235}U is a finite resource, if it continues to be consumed at the rate it is currently being consumed at, its life can be measured in decades, not centuries according to available sources. There is not enough of it. ^{235}U is a fairly rare isotope of Uranium. This isotope decays naturally and spontaneously to produce various fission products and quite a lot of heat. It is the production of heat in this process which is useful for the purposes of power generation. There are a number of different types of fission reactor. In all the different types of nuclear reactor however, power is produced by heating water to make it boil. The resulting steam is then used to drive turbines which produce electricity.

Job

One of my first jobs was working as a lab technician for the United Kingdom Atomic Energy Authority (UKAEA) at Winfrith in Dorset. Describing a few things about my experience there and about the project I was involved with will be helpful in explaining my conclusions about nuclear power.

I was attached to the experimental 'Dragon' High Temperature Gas Cooled (HTGC) reactor project. There were two main design factors which made the Dragon reactor different from other designs. The first was that it was cooled by Helium gas at high temperatures, 800^0-1500^0C. The second was that the uranium was contained in tiny ceramic pellets. The idea for the pellets was that if there was an overheating problem, and the uranium and/or the reactor melted, the uranium would be contained within the pellets and could not collect and form, what is called, a 'critical mass'. If ^{235}U reaches this 'critical mass' it will explode. Incidentally, that last sentence explains the second reason, or it may be the first, why the authorities value ^{235}U.

You can make it explode. You could make a weapon with it.

The two design factors, high temperatures and protective pellets, illustrate the inherent tensions in any nuclear reactor. To make them work, it is necessary to allow the temperature to rise. But if the temperature rises too much, the fuel can melt; indeed the whole reactor can melt and shed the fuel. If the fuel collects at the bottom of the reactor vessel, all kinds of horrors can happen.

A nuclear power station is powered by a controlled nuclear explosion.

The danger of an uncontrolled explosion is bad enough. But this is no ordinary explosion. An explosion that puts the contents of a nuclear reactor vessel into the atmosphere will, depending on which way the wind is blowing, make hundreds of square miles unfit for human habitation, unfit for agriculture, unfit for any human use.

The explosions at Fukushima in 2011 have made vast tracts of land around the power station uninhabitable and out of bounds. It is unclear whether the explosions at Fukushima were nuclear, or simply chemical. The authorities are very secretive. Chemical explosions are a real danger in out of control nuclear reactors. At the Three Mile Island nuclear emergency in Philadelphia in 1980, a bubble of hydrogen had collected at the top of the reactor vessel as a result of the meltdown. It was only luck which prevented air mixing with the hydrogen and lack of a spark which prevented the whole structure being destroyed in a chemical explosion. The result would have been devastating. Again, there is secrecy about it, but in the initial television broadcasts, it was said that one of the Fukushima reactors was run on MOX fuel. MOX is the name given to mixed oxide fuel (it sounds innocent, doesn't it – like mixed nuts and raisins – mmm nice!). MOX is not nice. It is a mixture of uranium and plutonium oxides. Plutonium is nasty stuff. As a heavy metal, like lead, it is highly toxic. In fact, as well as being the heaviest metal, it is the most toxic substance bio-chemically. Add to that that it emits lots of high energy Gamma radiation (along with all kinds of other nasty stuff, like Beta radiation and Alpha particles); it isn't something that you would like in your backyard. But this may be what the poor citizens of that part of Japan have in their, now uninhabitable, backyards.

Fear

Some people argue that there have been very few deaths as a result of nuclear power, and that all the worries about nuclear are unfounded.To me, that is complacent, callous, and simply wrong. One death is too many. The illnesses that have arisen as a result of exposure to unwanted radioactive substances are well documented[1]. The fear of nuclear power and its associated radiation hazards is a perfectly rational fear. What makes Gamma, Beta and Alpha radiation so particularly iniquitous is that you cannot see it. You might be being irradiated by radiation from fissile material, but you wouldn't know it unless you went round with a Geiger counter all the time. And I think that kind of fear is something that nobody should have to live with. The sound bite 'the only fear we have to fear is fear itself' is correct when it comes to nuclear power. The authorities should indeed fear fear itself since it is perfectly *rational* to be worried about a potential hazard that you cannot see.

Dragon

While I was working at Winfrith, I was asked to assist my boss with taking a sample from a leak of Helium gas which was occurring somewhere within the inner containment vessel. Two thick concrete domes surround and contain the Dragon Reactor's core. These were known as the inner and outer containment vessels. One morning, we walked the half mile or so from the engineering labs where we normally worked to the Dragon Reactor's control room, which was separated from the domes by a short walkway. At that stage I was not really sure what we were about to undertake. The reactor had been shut down six days previously. Shutting down had meant moving various control rods in or out of the reactor pile. Actually, the heat producing reaction does not ever stop completely because the natural decay process continues regardless of the position of the control rods, so 'shutdown' really means partial shutdown.

We were given clean coats and new radiation monitoring badges to wear, we had a quick look around the control room and at the instruments, and then we set off to enter the dragon. A lot of time has passed since then, but I can still vividly remember some things from that visit. First of all I remember the reactor building itself. I had not been up close to it until then. Close, now, it seemed bigger. It was a huge concrete cylinder, towering high above the four-storey control buildings below, making them look small. It was perhaps three or four times the height of those buildings. The low hum emanating from somewhere deep within the thing seemed to give it a presence. There was a door at the end of the connecting corridor. It was huge, a bit like the doors for bank vaults that feature often in bank robbery films, but two or three times thicker. It took some time to wind that door open. We entered a small room, a bit like a lift room. It took a similar amount of time to close that door behind us – perhaps twenty minutes. We were in the first air lock. As far as I remember, there were no communications now between us and the control room. This was well before the era of mobile phones and good telecoms. The hum from within had increased in volume, and it was warmer in there. On the other side of the lift room was another door with a large locking and unlocking mechanism.

This second door was not as thick as the first door, but still was extremely heavy. We opened it, went through into another small room, swung that door shut behind us, and wound up the mechanism to close it. It had a large locking arm on it – like a large version of those locking arms you get on some launderette washing machines. We were now locked in, in another air lock, inside the outer containment vessel, but not yet inside the inner containment vessel. The door into the inner containment area was now in front of us. This was another impressive vault-like door on immense hinges. The temperature had gone up, and the noise was louder. Nothing would have prepared me for what I was about to feel in the next fifteen minutes.

Before we started to open the last door, Brian, my boss described to me that he would point out once we were inside, a large nut, about two feet in diameter. He said that there was a neutron beam coming out horizontally from that nut. I must step over that beam, he said, and not walk through it. I'm still not sure, to this day, whether he was pulling my leg so to speak, with a nuclear physicist's version of 'go get me a tin of striped paint'. I wasn't too concerned about that at the time, and was ready to attempt to obey any instructions.

The first thing I felt when the door opened was air, hot air, penetrating up my nose, into my head, right into, it seemed, the inside of my skull. This was hotter, much, much hotter than the hottest sauna I have felt since. The skin on my face was stinging. The second thing was the noise, everything was shaking, so it seemed. There was noise coming from everywhere. Looking up it appeared we were in some kind of immense cathedral. I could see the dome of the roof hundreds of feet above us. In front of me Brian was pointing out the nut and beckoning me to hurry up.

I stepped over what I imagined would be where the neutron beam would be and followed Brian the hundred feet or so across to the middle of the inner containment where there was a structure comprised of all kinds of girders and pipes and cabling that was five- or six-storeys high with a flimsy staircase attached to it. Up the staircase we went. At that point, I learned later, we were directly over the core, which would have been about 50 feet below the level of the metal floor we alighted on as we entered the inner containment. The reason for most of the noise was that the coolant, helium gas, was still being pumped through the reactor. At a point about 100 feet above the floor, we attached some equipment to a pipe, opened a bleed nipple and collected a sample of the helium coolant, took some temperature, pressure and flow measurements and then we got the hell out of there as quick as we could. I think we had been inside the inner vessel about ten minutes.

Safety problem

I suppose the point of telling this anecdote is to emphasize the human element. There is always the possibility of human error. And things made by humans always breakdown eventually and so have to be repaired and/or replaced on a continual basis. As long as there are nuclear power stations in existence, there is a constant chance of someone, somewhere making a mistake that throws a spanner into the nuclear works. For much of the last decade, most of the nuclear power generators in the UK were shut down for maintenance reasons[2]. All kinds of machinery can get us into difficulties. A wind turbine could lose a blade, a solar panel could fall off a roof. The consequences of this kind of accident could, indeed, be tragic. However, the likelihood of an accident like this causing trauma on a scale similar to the human misery that a serious nuclear malfunction could cause is minute. It is the *potential* for disaster and the scale of the potential disaster that nuclear poses that is the safety problem. We know from experience that nuclear accidents are not only a theoretical potential. Real disasters like those at Chernobyl and Fukushima have already occurred and it is only through luck that these events did not cause many more deaths. There have been numerous near disasters – Three Mile Island in the US., Windscale in the UK., and probably others that we have not been told about. The Windscale fire was kept secret for more than a decade by the Government and the nuclear authorities. And the more nuclear power stations that are built, the greater the chance of another accident occurring somewhere. Next time it might be a full nuclear explosion near a major city. Do we want to allow the reckless, senseless capitalists to wait for that to happen before we shut them down?

Safe; Clean; Cheap?

If nuclear power did produce safe, clean, cheap and plentiful energy as its supporters claim, then, perhaps a lot more people would be all for it. The problem is that it does none of these things.

It is not safe.

Is it clean?

No. The waste products from nuclear power operations take decades to clean up. The Dragon reactor – a relatively small experimental reactor – is a case in point. It was closed down in the 1970s. Its decommissioning took over 30 years.

Is it cheap?

No. When all the costs are taken into account: extraction and separation of uranium, training, construction of 'safe' power stations, decommissioning of the power stations and storage and clean-up of the fission products, then nuclear power is the most expensive way of producing electricity.

Is there enough?

The authorities are very secretive about the available Uranium reserves, so it is difficult to say. Some reports say that there is enough ^{235}U reserve in the US for generation at the current level to continue for about 50 years. But the actual reserve may not be the key limiting factor. The 'West' – ie the US, the UK and supportive administrations prevent some countries – like Iran for example – from developing nuclear technology. The nuclear elite are worried that other countries, besides them, will develop a nuclear bomb industry. Isn't this the give away? The reason nuclear power is wanted is for none of the reasons above. It is wanted by the governments which already have nuclear power and nuclear bombs so that they can maintain their political and military dominance. An active nuclear power sector enables a nuclear weapons industry to exist relatively cheaply. Without a nuclear power industry, it would be almost impossible to train up enough nuclear bomb building experts to satisfy the warmongers. This is relatively clear when nuclear policy is examined. There is no claim that nuclear power could provide anywhere near all of the electricity we need, let alone all the energy.

Is nuclear fission necessary?

No, all our energy can come from renewables – nuclear fusion energy from the sun.

Summing Up

To start with then, despite claims to the contrary from some learned professors, it is very possible to produce all of California's and the UK's energy requirements using renewable sources alone. By extrapolation, because ambient energy is spread more or less evenly across the globe, it is possible for all people everywhere, with a suitable investment in infrastructure, to reduce to zero the use of fossil fuels.

The reason that the right infrastructure decisions are not being made is because a network of the world's richest and most profitable multinational corporations will not invest in energy efficiency or switch to renewables. Switching would leave their huge investment in fossil and nuclear energy stranded. They are determined to continue raking in profit in a market based on producing and selling climate damaging fossil fuels regardless of the effect this will have on the environment.

I will continue by making an assertion.

The over five thousand people who died, and the tens of thousands who were injured in November 2013 when the most powerful typhoon ever recorded to have hit land devasted millions of peoples lives in the Philippines and the twenty-seven who died in Oklahoma as a result of the tornado in May 2013 and the twelve hundred workers who died in the collapse of the Rana Plaza garment factory in Dhaka, Bangladesh in the same month all died for the same reasons – as a result of the decisions and actions of a few in the pursuit of profit.

Climate scientists can now, arguably, justifiably assert that the severity of the Philippines typhoon, the Oklahoma tornado, and recent hurricanes such as Hurricane Sandy, and the increasing numbers of extreme climate events all around the world, are due to anthropogenic global warming. The severity of these climate events does not have a natural cause. The extreme amount of fossil fuel burning which is continuing at an ever increasing rate at all places on the globe is leading to climate disruption. The decision to continue to invest in fossil fuels, and not in safe alternatives, is made by a tiny handful of super-rich energy industry tycoons, with the tacit approval of their respective governments. These decision makers are responsible for the deaths which result from their actions.

It may not be possible, statistically, to link individual weather events to any specific cause, but, 1) we have been warned by the climate scientists that it is very likely that the strength and frequency of storms will increase if the temperature of the earth's surface increases and 2) the vast majority of experts in the field agree that it is the increasing levels of carbon dioxide being released into the atmosphere from human actions that is causing the measured surface temperature increase.

Even if there were only a slim chance that the record breaking wind speeds that have been recorded in the most recent storms are being caused by behavior that can be changed, it would be incumbent on a sensible, non-reckless world leadership to enforce a change. In this case the change necessary is a rapid, planned decommissioning of the coal, gas, oil and associated industries. But, it is not a slim chance that the patterns of climate change we are seeing are down to human actions. If you read the recent reports of climate scientists, it is plain to see that it is a virtual certainty that the severity of the recent storms is the result of anthropogenic global warming. I think we can go further: the severe storms and the resultant deaths have come about as a result of profit induced global warming - the deaths have come about as a result of capitalism. If the essential changes are not made, and soon, there will be many more storms, floods and droughts. They will get more extreme, and more and more of the poorest and most vulnerable people on the planet will die. But no change is being enforced, and there is no prospect of the necessary change being enforced judging from the recent actions of the current world leadership.

Dirt

The garment workers died at their workplace because their employer treated them like dirt. What drove their employer to treat them like this? Was he simply a sadist who liked herding people into unsafe places just for the fun of it? No, his motive was, and still is, profit. The less of his capital spent on safe buildings, he calculated, the more would be his profit. The only reason why workplaces in the 'developed' countries are safer is because workers in the past have fought for this. Resistance works.

The few

Decisions under capitalism are made by the few for the benefit of the few. But the ruling class - the few - the enemy within, don't get away with doing just what they like. This is because of the resistance and solidarity of the many, because of the resistance of the working class. Where there is organized resistance at work there is safer working and better wages. The greater our solidarity, the safer we are. If we can increase working class solidarity using the lever of increased union membership, safer working conditions will follow. And, of course, workers are not only interested in safety at work. They also want a safe place to live. Worker organisation and resistance is an essential part of the campaign for a safer environment.

As discussed earlier, workers produce all the wealth socially. But their product is immediately taken off them. It is appropriated privately. A historical advance will have to come from moving to a situation of continued social production *combined with social appropriation*.

I don't know whether it will be possible, under the present, capitalist system, to prevent a climate disaster from occurring. But I think it is looking more and more unlikely. After all, it is already too late for some thousands of people. As has already been mentioned, you could say that capitalism has already failed the task. I think the struggle for a safer environment will take us towards a better system - a system where the needs of all are paramount, a system where production is for need not profit.

Decisions then, under a system of socialism, will be made by the many to satisfy the needs of the many. Production will be done on the basis of need. Need for safe working conditions, a clean environment, good food, housing, health and education, access to leisure, art and culture etc will trump any backward ideas of exploitation for profit.

Forming stronger union organisation will be a start towards this, because the stronger our voice, the more we can force the owners of production to concede this or that demand – the safer we can make our workplaces and the more we can access more of the product we produce The greater our strength and our successes in this endeavor, the more confident we will be to make further demands for a safe environment and for a living wage for all.

The first requirement in a society based on production for need will be the health and safety of the producers. Under the present capitalist society, unionization of work places is a first step in the right direction. Workers resistance can produce safer working conditions, and there is and must be a constant struggle to maintain health and safety. Under the capitalist system, for the capitalist, the health and safety of the producers is an unhelpful interference in the business of making a profit. In a society based on production for need - a socialist society - it is the first principle.

Record highs

As I was writing this, in mid 2013, it was announced that the atmospheric concentration of CO_2 was almost 400 ppm. The CO_2 concentration last reached this level over 4 million years ago.

I don't think that it is a coincidence, that in almost the same news broadcast, it was announced that the stock markets around the world had almost reached record highs. There is not a direct link, but there is a relationship between the amount of fuel burned and the activity in the economy.

However, even though world stocks were reaching record highs, the median wage in both the US and the UK economy was going down, adjusted for inflation. The austerity measures that are being implemented in an attempt to remove 'structural deficits' mean that people are working harder and harder for less and less pay. This is to be expected – it is the way capitalism works. The conservatives want to conserve their financial position. Anything that endangers this – such as people getting too much 'free' health service, or too many benefits, or higher pay, anything that means the rich and greedy might lose a little of their stash, means they will fight tooth and nail against it to achieve their ends. This is why the austerity measures have been imposed. They want to make the working class pay for what they call the deficit. As argued earlier, there is only a deficit if you accept that the ruling class may continue to make profits in the same old way and continue to get away with paying taxes at special low rates and in some cases no taxes at all.

Printing money

Their system is in crisis. For the last five years, since the crash of Lehman's bank in 2008, governments have been pumping money into the system, ostensibly in a bid to restore stability. They call this pumping, 'Quantitative Easing' (QE). QE has been going to banks (the very institutions responsible for the crash!) who lend it out to investors who have invested it in stocks which have then risen to high levels. When the stock market indices fall, as they have again recently, this means that the investors have sold, have 'taken their profits'. It looks as though the Central Banks have been, indirectly, pumping money into the pockets of these investors.

The worldwide number of 'high net worth individuals'[1] climbed by 10% in 2012[2] to 12million. In that year, the assets of these people increased to $46.2T (£29.5T), up 10% on the previous year. The total wealth of the UK's 500,000 'high net worth individuals' *went up* by £123B between 2011 and 2012. Coincidently, from March 2009 to July 2012, total QE for the UK amounted to £375B. That's roughly £125B per year of QE on average. Their wealth increased by £123B in 2011 and QE was almost the same (£125B) in that year. These figures do not include the value of homes or any 'consumer durables', (such as cars, boats, planes, second homes etc.

So, while there was a worldwide recession, the wealth of the UK ruling class increased. And, interestingly, it increased by almost exactly the same amount as was pumped in to the system by QE. The US ruling class fared just as well.

At the same time, indeed since the crash of 2008, wages have failed to keep up with prices, people have lost their jobs, and some their homes and livelihoods. 'We are all in it together', the mantra of many of the world's ruling class, referring to the current 'austerity measures', could not be more wrong. In fact, the austerity measures are for us only. For them, prosperity measures apply. It is prosperity for them, austerity for us.

Actually, those high net worth types referred to above are rather poor compared to a more select group (of 110,000 people worldwide) who are even richer. On average, these 'ultra high net worth individuals' are thirty times, or more, richer than those discussed above.

Where did the money that went into the pockets of the rich come from? Did they work for it? No. It came directly from us. As our wages fell, their wallets got fatter.

Money for nothing

I am not an economist[3], and I don't pretend to understand

exactly what they are doing and precisely what they are trying to achieve with this 'creation' of money, but, I say we shouldn't let them fool us with it. They want to make sure their nests are feathered. They are not too concerned about us as long as we keep quiet. At this point in the economic cycle (of booms and slumps), the printing of money doesn't seem to be creating jobs and this is probably because, as discussed before, investors cannot see a way of making a profit from the exploitation of human labor: the risk is too high; the rate of profit is too low.

You cannot create wealth from nothing; by, for example, printing money. Wealth comes from work. Although people want jobs, the system is not allowing the creation of jobs, even though there is a desperate need for all kinds of things to be done – houses built, medicines distributed, wind farms constructed, flood barriers constructed, teachers trained etc.

When the capitalists print money and it ends up in their bank accounts, they are simply robbing us. Yes, banks can 'create' money - by lending it. But there must be a demand for that money. Capitalists might borrow money if they see the possibility of converting it into capital by exploiting human labour (employing workers to produce commodities). But if there is a slump, capitalists will not borrow - there will be no demand for credit. Ultimately, under the capitalist system, surplus value (profit, wealth for the boss) and wages etc can only be created when workers are exploited.

All the money making schemes which the capitalists dream up - for instance, 'offshore' tax havens (based in places like the City of London or the Cayman Islands) where they secrete their loot, hiding it from the tax authorities, obscure trusts holding various financial asset classes, accounting tricks to create 'structured investment vehicles' etc, all, at the end of the day, are funded by the exploitation of workers[4].

Contradictions

Capitalism is contradictory. It creates plenty and want in unequal measures. There is tremendous progress alongside barbarous destruction.

Because of the way it has organized workers in a highly productive, systematic, planned division of labor, previously unachievable (for instance compared to the feudal system of production) it has been able to generate wealth unimaginable to previous societies.

In the past four hundred years, since the coming of capitalism, production of all manner of things has expanded in an exponential fashion.

But the result is plain to see. There is enormous wealth alongside devastating poverty.

Advanced scientific discoveries have both eliminated killer diseases and also armed brutal wars. History moves forwards and backwards at the same time. We understand much more about nature and ecosystems than our ancestors ever did, but the systems we live under are causing irreversible loss of species and bio-diversity at an alarming rate.

Biodiversity

And of course, the bio-diversity, is not just a 'nice to have' option it is essential for human life. Some eco-warriors attempt to explain the importance of bio-diversity through the concept of 'eco-system services' – services that the environment provides for us for no payment, which we take for granted. Environmental economists have worked out the 'value' in Pounds, Euros or Dollars, that various aspects of the environment deliver. They work out, for instance, how much it would cost to pollinate a crop if there were no bees or insects to do it. In that way they put a price on bees etc. While this kind of analysis is an attempt at highlighting the 'value' of this or that part of the environment to human life, the analysis, because it conflates use-value with exchange-value, also allows for these parts of the eco-system to be made into commodities that can, of course, be bought and sold. Already the British, Tory-led, government is talking about 'environmental offsets' where developers are allowed to destroy a certain piece of land or forest, or stream, or other habitat etc., as long as they provide an alternative.

Marx

The environment or the ecosystem we live in is, of course, not simply a supplier of services. It is fundamental to us, and to all life. Karl Marx, one hundred and fifty years ago, showed, as part of his critique of capitalism, how we are one with our environment, in a quite fundamental way. He talked of what are now described in terms of 'ecosystems' and 'the environment' as our **'inorganic body'**[5]. By that he meant that we were, in the same way as we are dependent on our heart or our brain or our spleen, or our bones, dependent on the earth and the plants and animals that it supports for our existence on the planet.

The relationship is, of course a reciprocal one. But, under capitalism, we have not been keeping our side of the bargain. Marx describes how in his time, a 'rift' had developed and was developing. Agricultural methods that came into use under capitalism were depleting the soil, making it less and less fertile. The cities and the people in them were becoming separated from the land. Nutrients were taken from the land, but not returned. In fact the nutrients that earlier would have been returned to the soil, were now being disposed of in ways that would not only result in less fertile soils, but would degrade the rivers that could be a supply of clean water. Bluntly, the nutrients in human urine and shit, nitrogen, potassium, phosphorous and other micro-nutrients essential for healthy plant growth were, and still are, being wasted and flushed away down the tubes while other resources were and are used to produce fertilizers artificially.

Since Marx's time environmental degradation has accelerated.

As Chris Harman has described[6], capitalism is digging up the very ground that it is standing on. Using Marx's analogy, we are cutting through of one of the vital parts of our body. The capitalists are chewing on our living organs. Whatever system follows capitalism, these cuts will leave scars. Species loss is not reversible.

Capitalism has reached its use by date. It reached it some time ago. Arguably, it has had its uses, it has been useful for humanity. As Alex Callinicos has written, 'Capitalism is a necessary pre-requisite for Socialism'[7]. It has developed the forces of production to the point where enough is produced to satisfy the needs of all. The trouble is it does not satisfy the needs of all. This is because it only attempts to satisfy the greed of the few.

Capitalism has created the material conditions for its own destruction. It has created a massive working class. It is now possible, nay, necessary, for the working class to seize the levers of production, developed under capitalism, to make a world fit for all to live in. If we don't, the capitalist class, in pursuit of ever more profit, may well drive all of us over an ecological cliff from which there might be no return.

Act

So, what is to be done, what can we do?

Firstly, as argued previously, form and join unions at work. Build solidarity with other unions, communities, and other workers organizations right across the economy. Form links with all the groups of people fighting the austerity cuts – pensioners' action groups, the unemployed, students fighting fees, etc. This is not an easy task, but it must be done. Our history provides a guide as to how it can be done; it shows it can and must be done as part of the struggle against our oppressors and exploiters. However, that cannot be enough. The ruling class acts as a class to defend its interests. It produces laws, hires police forces, and hires all kinds of agents to look after its interests. How will the working class *act* as a class? That is the question!

In this case the answer is neither 'to be' nor 'not to be', it is to act.

The working class becomes conscious of itself as a class when it acts as a class.

The ruling class attempts to prevent this happening and uses every weapon in its armory to achieve this. Principally, it uses divide and rule tactics – racism, sexism, nationalism to divide our side. These tactics tend to fragment the working class, creating unevenness in the level of consciousness. But it is in the process of fighting and defeating racism, of fighting and defeating sexism and homophobia, of fighting the lies of nationalism, of fighting together for better wages and working conditions that a strong unified class-conscious working class can be forged. In the process of fighting for better conditions, it will be necessary, *in order to win,* for backward ideas such as racism, sexism, homophobia and nationalism to be smashed and thrown aside by a victorious working class.

The working class everywhere must find a way of using its greatest strengths – its great numbers, its immense intellectual resource and creativity and solidarity on every level - from local workplaces to community groups and vice versa through to international solidarity. The capitalist class works on an international basis. So must we.

Thread

As you might have gathered, I consider myself to be a Marxist. But that is not all. I also consider myself to be an Engelsist, a Trotskyist, a Luxemburgist, a Leninist. Marx and Engels, Trotsky, Luxemburg, Lenin and many others fought and wrote about workers struggles in the past. For us to see a clear way ahead, we need to be able to learn from the past. We must use the method of Marx, Luxemburg, Lenin and other fighters to analyse where we are in the struggle. We are lucky that we have giants, like

Marx, whose shoulders we can stand on to enable us to see more clearly the line we need to follow.

People might say that the crimes of Stalin against the Russian workers and the record of the USSR proves that the Bolsheviks, who followed the ideas and actions, the red thread of Karl Marx, Frederick Engels, Rosa Luxemburg, Vladimir Lenin and Leon Trotsky, were wrong. I think this view (incidentally, a similar view was held by both the state-capitalist, Stalinist leadership of Russia and also the capitalist leadership of the West) is itself completely wrong. Stalin had Trotsky and many of the Bolshevik leadership murdered, and the Western leadership considered Trotsky and his followers to be enemies. My view is that, it wasn't Bolshevism that led to the defeat of the working class in the revolutionary upheavals of the early part of the twentieth century, but a *lack* of Bolshevism. For the Russian revolution to have been ultimately successful, the ideas of Bolshevism needed to spread beyond the borders of Russia. But, the Bolshevik parties outside of Russia were either non-existent, or young and unproven. The problem was not Bolshevism, but that there was not enough Bolshevism.

It is important, I think, to follow Marx's motto – and doubt everything, and to not slavishly follow leaders. Leaders, like everybody else, can get things wrong. Trotsky thought that, after the Second World War, Russia would be able to revert to the path of socialism. He was wrong. State Capitalist Russia, trampled on its workers in order to compete militarily with the West. We can stand on the shoulders of giants and see further. But we can do so, only if we keep our eyes open. The revolutionary socialist and Trotskyist, Tony Cliff, first argued, in 1949[8], that Russia was not, as Trotsky maintained, a workers' state, but was State Capitalist in character. This insight has been terribly important for the 'left'. It has allowed the 'red thread' to continue, allowed the green shoots, of revolutionary socialism to grow, and left the forest of Stalinism dead and dying.

Cliff

Revolutionaries like Tony Cliff, and many of his friends, such as Duncan Hallas, Paul Foot, and Chris Harman have continued the revolutionary socialist tradition, have strengthened the red thread, and nurtured the roots of revolutionary socialism – of 'socialism from below', of a socialism capable of smashing capitalism.

I believe people who want a change to the system they live under will benefit from closely studying what these fighters were saying. They are our modern giants whose shoulders we can stand on in order to determine the best line of attack.

Marx and Engels changed their ideas as they learned from the struggle of their day. For instance, it was not until the Paris Commune of 1871 that Marx was clear about whether workers could take over the state or not. The experience of the commune showed that the capitalist state could not simply be taken over. The workers had to smash and break up that state and form one in their own image.

Similarly, we must continue to learn from our struggles.

Twice in the last forty years, UK prime ministers have been ousted as a result of workers struggle. The first time was in the early 1970s. Ted Heath, Tory prime minister in the early 1970s was forced to call an election under the slogan 'Who rules the country - the unions or the government?' On the 10th of February 1972 tens of thousands of engineering workers and other workers went on strike and walked out of their work places all around Birmingham and marched to support the miners who had been attempting to close a coking plant as part of their struggle for better wages from the coal bosses. The result was a spectacular success. On their own, the miners had been struggling to close the gates at the Saltley coking plant, because of the number of police employed to break the strike. The police had been able to keep the gates open for the previous week. But when ten thousand engineers marched over the hill down to the plant, the chief of police called the prime minister to tell him that he would have to close and lock the gates of the coking plant, otherwise he would not have been able to guarantee safety in Birmingham. The combined workers action closed the gates. The battle of Saltley Gates has gone down as one of the working classes greatest victories. It demonstrated the power of the organized workers when they use their strength - their number, and solidarity. It wasn't simply a win for the miners. The whole working class benefited from their victory. As one of the pickets said - 'The feeling of power and unity we had was indescribable. You just felt so elated because of the power that the workers had shown'[9].

Workers confidence in their power rose over the next two years as was shown by a rising level of strikes and occupations and other actions which forced Heath to put the whole country on to the three day week and then to call an election. Heath's Tory government lost the subsequent election.

A second spectacular 'win' for the working class, which exhibited workers power in action, occurred in the late 1980s and early 1990s. The then Tory flagship tax - the Poll Tax of Prime Minister Margaret Thatcher was defeated by a sustained campaign which lasted over two years. In the end, the campaign against the Poll Tax not only removed the tax, it brought down and defeated the Tory Prime Minister.

Although the anti poll tax campaign was not fought in the workplace, the overwhelming number of those in the campaign were workers and unemployed workers and members of working class communities. Workers from a myriad of different work places combined together in small and large places up and down the country with the sole purpose of defeating the hated Poll tax.

Tony Cliff, national secretary of the Socialist Workers Party (UK) at the time joked once that he loved the Poll Tax. This was, he explained, because it took on everyone at the same time – all workers. The Tories had miscalculated; they had forgotten that they need to divide our side in order to win.

One little anecdote. On the morning of 31st March 1990, the day of the great anti poll tax demonstration in London which ended in a police induced riot, I was on one of the seventeen coaches which had traveled the 160 miles or so down from Sheffield. There was a traffic jam as our coach entered inner London. It was our good fortune to observe a very amusing incident - an incident very probably provoked by our presence. Between two of the coaches in the jam there was a car. The car was a Rolls Royce. The well dressed people in it had been laughing at us. Anybody who drives a Rolls Royce is, really, 'giving the two fingers' to the working class. It so happened that there were a group of road workers cutting up the pavement (side-walk) with a circular stone cutter on the opposite side to the car. On hearing our 'No Poll tax, No Poll tax' chant, one of the group of workers, cheered on by all of the surrounding coaches and his mates, took the scary looking power cutter tool right up to the Rolls Royce, and made a gesture as if he was going to cut the car in half. The look on their faces of the people in that car was then a sight to behold. Apologies to the chauffeur! Of course, that time, the stone cutter didn't cut the car or hurt anyone! It was a very small but very interesting spontaneous display of workers solidarity.

The Financial Times is a newspaper that tells the truth from the viewpoint of the ruling class. This is because investors need somewhere they can find out the facts about what is happening in their world in order that they can invest their unearned wealth where it can make them more profit. The FT provides that service. It made my day when, after Thatcher had gone, and after the Poll Tax was consigned to the dustbin of history, I saw a photograph in the Financial Times. The photograph showed a picture of someone carrying a Socialist Worker placard which read 'Stuff the Poll Tax', and under the photograph was the simple caption 'And they did!' The poll tax was defeated by a sustained campaign by workers up and down the country.

Although the current austerity measures, in the UK and everywhere, are much more severe than the Poll Tax, similarly to that tax, these austerity cuts attack everyone – all working class people. And we know that now, and in the future, the ruling class will continue to attack us and try and make us pay for their lavish lifestyles and their system, by grinding our side down with their laws and lies. But we do not need to look too far into our history to see that, if we use all our power and we are united, we can win. I think the answer for how we maximize our power, how we can win, can be glimpsed in the two types of victory outlined above.

Primarily, our core power is *in* our workplace - the power glimpsed at Saltley Gates. But we don't always have the confidence to use that power. What that fight showed is that, even the strongest union of workers - for example the mine workers - can not win on its own. But we are strong and can beat any capitalist when we fight together. And, if that sort of workers power - the sort exhibited at Saltley Gates is combined with a sustained campaign like that against the Poll Tax - a bottom up, nationwide campaign involving the widest layers of the working class, we can defeat the most determined government onslaught.

I think it might help us, if we dare to imagine what our future revolutionary workers councils might look like (ie how a successful fight against the system itself might shape up). In the UK, I think they would bear some resemblance to the grass roots organisation that started to form towards the end of the anti poll tax campaign. In Sheffield for instance, a diversity of anti poll tax groups from around the city sent delegates to a city wide organisation which discussed and decided on forms of action. Similar ways of building and organizing happened up and down the country. The anti poll tax campaign was built around a non bureaucratic bottom up campaign of loosely affiliated groups of workers and community groups, together with a national focus. That sort of framework, combined with sustained strikes, demonstrations, occupations and picket lines will defeat the most organized ruling class. And it is important to remember that the ruling class makes mistakes and is prone to fracturing. In all countries, we can learn from victories in our own and in other countries. For example, Egyptian workers attempting to win a better world, to progress their revolution can learn from the struggles in Greece and vice versa etc. We can all gain from victories everywhere and learn the lessons.

The British State is not invincible, as was revealed during the anti-poll tax campaign and at Saltley Gates. The Government can be defeated, but it will try to use all the powers at its disposal to defeat its adversaries. For that reason, we must be able to match the government symmetrically, to maximize the strength and numbers on our side, and directly oppose their attacks, all the austerity measures of the government, if we are to win.

I live in the UK, and the examples I know most about are obviously in the UK. All around the world though, in every country, on every continent, people are facing the same sort of challenges. The capitalists are trying to make us, the workers, pay for the crisis of the system they uphold. The answer is the same everywhere: Unite and fight.

They will try to divide us because they know unity is our strength. But we have an answer for them. Workers of the world unite! was Marx's famous phrase and it is still good for us today.

We must look to what unites us.

It is hard and sometimes thankless work fighting the system, and inevitably people get tired! We are human and, on our own, we are weak but, together we are strong and I urge those who have let go of 'the red thread' to grasp it again, to rejoin it, to strengthen it. And I urge workers everywhere fighting back to unite with other workers, to grasp the thread. Together we can make a red rope with which to hang capitalism. The working class, the workers party, needs you. And if you want to win change, you need it. The clock is ticking for the climate. But I think a victorious working class can lead us to a safe planet. We have the time and we have the power. What we have to do now is to organize and use it! As Trotsky said, in order to learn how to ride a horse, you have to get on it.

I urge workers everywhere to get on the horse.

A change is possible now, and ever more necessary.

Notes

Introduction

1 See Part 2

Part 1

Chapter 1
Ambient Energy, Intermittency and the Electricity Grid

1 Intergovernmental Panel on Climate Change, Climate Change 2007: The Physical Science basis, 2007, (Cambridge University Press, Cambridge

2 Boyle, G, 2004, Renewable Energy, Power for a Sustainable Future, Oxford University Press, p291;

3 Digest of UK Energy Statistics (DUKES) 2007
(Total Energy Consumption = 155Mtoe [million tonnes of oil equivalent] = 1856TWh)
Electricity consumption was 29.5Mtoe = 350TWh

4 National Renewable Energy Laboratory (NREL) April 2011 http://www.windpoweringamerica.gov/docs/wind_potential.xls
(nb this url downloads the xls file)

5 US Energy Information Administration, 2009 http://www.eia.gov/state/seds/hf.jsp?incfile=sep_sum/html/sum_btu_1.html
Total US Energy Consumption 94,446.9 Trillion Btu (= 27,682 TWh)

6 Boyle, G, Renewable Energy, Power for a Sustainable Future, Oxford University Press, 2004, p98

7 http://www.cleantechblog.com/2011/10/what-if-every-residential-home-in-the-u-s-had-a-solar-rooftop.html
This study reckoned PV output from total US roof space = 1,424TWh,
Total US Electricity consumption (2009) = 3,953TWh

8 Environmental Data Services (ENDS) Report 399, April 2008
9 Boyle, G (ed), 2007, Renewable Electricity and the Grid, Earthscan, p38
10 op cit
11 Daily Telegraph September 17 2011 http://www.telegraph.co.uk/earth/energy/windpower/8770937/Wind-farm-paid-1.2-million-to-produce-no-electricity.html
12 See Kempton, W. and Letendre, S, 1997, Electric Vehicles as a new Power Source for Electric Utilities, University of Delaware, USA
13 Smith Electric Vehicles http://smithelectric.com/
The world's leading manufacturer of commercial electric vehicles.
14 There are several electrical storage technologies commercially available.
For examples see Bent Sorensen, Renewable Energy, 2004, Elsevier Academic Press pp523 or Twiddell, J. and Weir,T, 2006, Renewable Energy Resources, Taylor and Francis, pp489
15 See Chapter 3

Chapter 2
Hot Air

1 University of Edinburgh economics professor Gordon Hughes in a report for the Global Warming Policy Foundation, which is chaired by former British Conservative Chancellor Nigel Lawson, reported in The Australian March 9, 2012
2 Australian citizen being interviewed, talking about wind turbines on BBC Radio 4 'PM' programme, 17/4/2012
3 Daily Telegraph September 17 2011
4 Prince Frederik of Denmark at the European Wind Energy Association (EWEA) annual conference, April 2012

5 The Guardian, Monday 29 June 2009
6 Bloomberg New Energy Finance, 10 November 2011 http://bnef.com/PressReleases/view/172
7 Professor Dieter Helm BBC 15 Aug 2011
8 See for example this article. http://www.socialistworker.co.uk/article.php?article_id=10073 http://preview.tinyurl.com/btbr8me
9 Quote from Eleanor Marx See www.marxists.org
10 Guardian http://www.monbiot.com/2005/12/05/the-struggle-against-ourselves/
When George writes in the Guardian article of Dec 5th 2005: 'The notion that we can achieve this by replacing fossil fuels with ambient energy is a fantasy', he is assisting, only, the very fossil fuel industry he seems to want to oppose.
And when he writes: 'Fossil fuels helped us fight wars of a horror never contemplated before, but they also reduced the need for war', he is forgetting that the wars that he opposes – for instance, the Gulf wars, were fought over the control of oil and oil profits.
Thirdly though, when he writes:
'Ours are the most fortunate generations that ever will [live]', he is telling us he cannot imagine a society more advanced than the one we currently live within.
This is where I differ. I think he is profoundly wrong about that. It is possible to have a world of abundant energy, food, shelter, recreation, art, fun etc for all. The solution though, is not simply technical, but profoundly political.
11 http://www.youtube.com/watch?v=aIBZNp8Znlc&list=FLbYk-gR34seUTdFvTOESthA&index=3&feature=plpp_video http://preview.tinyurl.com/7j9zqxb
12 see note 10
13 http://www.monbiot.com/2012/03/15/press-release-why-we-have-written-to-cameron/
Please see part 11 for arguments about nuclear technology.

14 see for example, T. Markvart, Solar Electricity, p1.

15. Actually, that statement is not strictly correct because the earth's energy balance has been disturbed by anthropogenic CO_2 emissions.

The earth is continually receiving energy from the sun and emitting energy back to space. If less energy is re-emitted back to space than is being received, the temperature of the planet will rise. It will continue to rise until an equilibrium position is reached where the incoming energy from the sun is matched by the reflected and re-emitted energy.

So, in fact, as we know now, the earth is not currently in equilibrium – the temperature is rising. This is because more (infra-red) radiation is being absorbed by the atmosphere and re-emitted back to the planet's surface than was the case before human activity caused the carbon dioxide concentrations in the atmosphere to rise.

The earth's temperature is being forced up. (This effect is called 'Radiative Forcing').

Unless we stop the increase in carbon dioxide concentrations, and start to reduce them, the temperature will keep on rising until another equilibrium state is reached where the earth's energy in/energy out equation is balanced again – but this time, one where the earth's temperature is higher than before.

The science of global warming has been written about extensively elsewhere*and it is beyond the scope of this book to cover this in any more detail.

[*See for example IPCC Physical Science basis 2007 or ZeroCarbonBritain 2007]

16 For a discussion on biological conversion and storage of energy, see Bent Sorensen, Renewable Energy, 2004, Elsevier Academic Press, p 289

17 Emeritus Professor of Energy Conversion, Ian Fells of Newcastle University [BBC News 17 February 2012]

18 Sustainable Energy-Without The Hot Air p169
19 See Appendix.
20 http://www.withouthotair.com/download.html
21 Sustainable Energy-Without The Hot Air p33
22 Sustainable Energy-Without The Hot Air p29
23 see op cit p29 and UK Gov statistics
24 UK Gov statistics
25 See Mackay op cit p 263 for explanation of how a wind speed of 6 metres per second equates to a power density of 2W/m2
26 op cit p265
27 See Appendix
28 op cit on p265
29 See for example Mathew, S., 2006, Wind Energy, Springer, p11
30 http://www.whiteleewindfarm.co.uk website
31 appendix
32 http://www.whiteleewindfarm.co.uk website
33 Carnedd Wen Wind Farm and Habitat Restoration Project, Volume 5 (Non Technical Summary), RWE npower renewables, September 2011
34 See Chapter 3 – Britain
35 John Humphries BBC Today programme 16/5/2011
36 United Kingdom Energy Research Council (UKERC) spokesperson BBC Radio 4 9/10/2009

Chapter 3
Powering the World with Renewables

1 See the IPCC report 2007
2 UK public data about energy use gives values for 'Primary Energy' and also for 'Final Energy Consumption'. The 'primary' energy figure relates to the total energy released when a fuel is consumed. Approximately one third of the primary energy is unused and is lost as waste heat. The final energy consumption is a measure of how much energy is

consumed as useful energy by final end users. The Digest of UK Energy Statistics for 2012 figures (DUKES) for 'Primary' and 'Final consumption' energy are 2820 TWh and 1961 TWh respectively.

See https://www.gov.uk/government/publications/digest-of-united-kingdom-energy-statistics-2012-internet-content-only

3 From the Digest of UK Energy Statistics (DUKES) for 2007

4 This is approximately 8.7 metres per second. The figure given for the wind speed is actually an average of the anemometer readings taken over one hour.

5 Divide the annual energy consumption by the number of hours in a year.

6 See Appendix

7 I assumed electrical energy losses in the power grid of 7%, in the power equipment (inverters etc) of 10% and in the storage medium, of 12%. See (http://www.eia.gov), (http://www.mpoweruk.com/energy_efficiency.htm) and (http://www.teslamotors.com/goelectric/efficiency) respectively.

Incidentally, D Mackay suggests a figure of 1% may be achievable for cable losses, op cit p 104.

8 The storage capacity of Dinorwig is 9.1GWh. The energy production difference between an hour of 19 knot winds and an hour of 17 knot winds at the single wind farm is about 25GWh

9 'A major R&D effort on energy storage and storage systems will be crucial for the achievement of a low-carbon energy system' [Stern. N, 2006 p255] (The Stern Review)

'Electricity generation from fossil fuels with carbon capture and storage will ... be unable to enter the transport markets unless improved and lower cost forms of hydrogen storage or new battery technology are developed' [Stern. N, 2006 p 423]

'The immediate conclusion is that until new solutions emerge that will add substantially to the overall capacity credit of a more varied combination of variable energy sources, perhaps including very substantial energy storage capacities, much otherwise uneconomic conventional plant will need to be retained or replaced' Laughton, M., 2007, 'Variable Renewables and the Grid: An Overview', in Boyle, G.(ed) Renewable Energy and the Grid, Earthscan, Chapter 1, p28

'if cheap and effective storage were to become available, it would be widely used in electricity generation systems ... if the storage could be small and distributed, this would have the added benefit of reducing the capacity requirements of both the transmission and distribution systems' Infield, D., Watson, S., 2007, 'Planning for Variability in the longer term: The challenge of a truly sustainable energy system', in Boyle, G.(ed) Renewable Energy and the Grid, Earthscan, Chapter 11, p 201

10 See eg http://www.teslamotors.com/goelectric/efficiency

11 See for example Mackay, D (2008) Sustainable energy without the hot air. UIT Cambridge, p 127

12 See below

13 Vehicle to Grid power was first described by Kempton, W. and Letendre in Kempton, W. and Letendre, S.,1997, Electric Vehicles as a New Power Source for Electric Utilities, University of Delaware, USA.

See also my MSc Thesis: from the Graduate School of the Environment (Director Mike Thompson) http://roar.uel.ac.uk /600/

14 354/8760TWh = (hypothetical) constant electrical power demand = 40GW. Each vehicle would contribute $40/32.5 \times 10^6$ GW = 1.2 kW. At 240v, the current would be around 5 Amps

15 See Boyle, G. (2007) Renewable electricity and the grid. Earthscan

16 Excel spreadsheet
17 Supplied by the Met Office
18 See Tables A2 and A3 in the Appendix
19 See http://zerocarbonbritain.org/index.php/report July 2013
20 See for example the One Million Climate Jobs report, produced by the Campaign Against Climate Change Trade Union Group (part of the Campaign Against Climate Change) http://www.climate-change-jobs.org/about
President Obama's energy minister, Professor Chou, has stated that 'energy efficiency is as important as CO_2 emissions'
21 See http://www.passivhaus.org.uk/
22 Waste deserves to be the subject of a book in itself.
Obsolescence, together with psychological obsolescence is built in. People are duped into buying stuff they don't need. Products are made to break down.
More and more garbage is produced year by year and capitalism, in the process, destroys the environment at an increasing pace.
Much of the waste is in the production process – packaging, advertising etc.
23 For example, Zerocarbonbritain2030
24 One million climate jobs now – op cit
25 See Kempton and Letendre (1997) op cit
AC Propulsion is a firm specialising in this technology: http://www.greencarcongress.com/2011/07/acp-20110713.html
26 www.energyalmanac.ca.gov and www.ecdms.energy.ca.gov
27 NASA Scientist James Hansen, is one of the more outspoken scientists linking recent extreme weather events to human induced global warming. See http://edition.cnn.com/2012/08/05/us/climate-change
The IPCC has published a report regarding the ongoing risks of extreme weather events: http://www.ipcc-wg2.

gov/SREX/images/uploads/SREX-SPMbrochure_ FINAL.pdf

Chapter 4
The Enemy Within

1 The 2009 report, 'Decision Time' is no longer on the CBI website. I found it here (with help from Friends of the Earth): http://docs.cumbriawindwatch.co.uk/Wind%20Docs/Cost%20of%20renewables%20&%20jobs/POLICY%20-%20CBI%20Decision%20Time%20PDF.pdf
I also have a copy.

2 For most signatories of The Kyoto Protocol, 1990 is the base year from which any binding reductions are to be made. Kyoto is intended to cut global emissions of greenhouse gases.

3 Emissions from cement production arise from two sources – from the burning of hydrocarbons to produce heat, and from CO_2 driven off from the rocks which are the prime raw materials used in its manufacture.

4 See http://www.esrl.noaa.gov/gmd/ccgg/trends/
http://co2now.org/current-co2/co2-now/
The CO_2 ppm readings are now taken daily. It reached the psychologically important level of 400 ppm for a reading in May 2013. The 'safe' level is 350ppm, see http://arxiv.org/ftp/arxiv/papers/0804/0804.1126.pdf

5 Only yesterday (20 July 2013) on the main British news channel, two prominent politicians made assertions about energy. 'The shale gas revolution is the best news on the energy front for a long time', 'there is no environmental problem at all [with fracking for shale gas]' – Nigel Lawson, ex British Chancellor of the Exchequer.
'There are no alternatives to nuclear and fracking' Paddy Ashdown, former leader of the Liberal Democrats.
Earlier in the week, the current Chancellor announced that the taxes on production of Shale gas would be cut by a half.

6 It's difficult to get information on the cross links between industries, but there is a long history of close relationships and joint ventures between mutually dependent companies (with the approval, tacit or otherwise of their respective governments).

So, for instance, Dr Beeching, the cabinet minister responsible for the closure of a great number of UK rail routes in the late 1950s was previously technical director at ICI, a company with strong connections to the oil industry. Auto and oil both benefited from the move from rail to road.

At the time the London trams were wound up in the 1950s, the director of London Transport was heavily involved with Leyland (autos) and AEC (Associated Engineering Company – a diesel engine manufacturer). With few moving parts, electric trams lasted forever – and they were therefore not a very good source for profits. And they didn't run on oil.

General motors and other companies were probably behind the demise of the electric street car in the US. For example see http://www.culturechange.org/issue10/taken-for-a-ride.htm

Dana Deasy GM North America CIO was Hired by BP to be its CIO and group president of its IT function (Digital and communications Technology DCT).

7 For example Exxon Mobil donated £1 million to organisations campaigning against controls on greenhouse gas emissions – City Am 19/7/10, p2

8 See Guardian report http://www.guardian.co.uk/world/2010/jun/07/bhopal-disaster-india-seven-convicted

9 See The Guardian, op cit

10 Hydraulic fracturing of rocks, using chemicals, to produce shale gas

11 See BBC Report http://www.bbc.co.uk/news/business-11497070

12 The main reason that the US is interested at all in the region

is the oil, and it needs and therefore supports states like Israel to act as its watchdog. Without massive American aid Israel would be incapable of acting in the way it does. See Phil Marshall, Intifada (1989)

13 See http://infochangeindia.org/environment/news/bhopal-gas-disaster-court-finds-union-carbide-chairman-officials-guilty.html

14 The price a commodity sells at is related to the amount of labor embodied in its production and market conditions at the time of sale. And see chapter 5.

15 Marx, K. 'Wages Price and Profit', Peking, p58

16 CBI report Decision Time (2009) op cit p22

17 The Feed in Tariff (FIT) was introduced to encourage take up of technology like photovoltaic panels (PV).

18 Energy Recovered Over Energy Invested (EROEI)

19 For a pessimistic view of the effects of a decreasing EROEI, see: http://www.tullettprebon.com/Documents/strategyinsights/TPSI_009_Perfect_Storm_009.pdf

20 See eg http://www.earthstructures.co.uk/retro.htm

21 See eg http://info.cat.org.uk/questions/wise/how-was-rammed-earth-used-wise

22 The farm at Wakelyns in Suffolk developed by Martin Wolfe shows the kind of agriculture which is possible, see eg http://www.organicresearchcentre.com/manage/authincludes/article_uploads/Martin%20Wolfe.pdf
Also see Simon Fairlea, Meat - A Benign Extravagance (2010), Patrick Whitefield, The Earth Care Manual (2004)

Chapter 5
Identifying the Agent for change

1 Rosa Luxemburg – On the Spartacus Programme

2 In 2007, Richard Portes, CBE, Professor of Economics at the London Business School co-authored a report on the Icelandic banks. The report asserted that Iceland's financial

regulator was of a 'highly professional' quality and that the banks were stable.

Frederic Mishkin then of Columbia Business school, and former Governor of the US Federal Reserve co-authored a report in 2006 giving Iceland and its institutions a clean bill of health.

In October 2008, Iceland and its banking system collapsed.

3 The meaning of Marx's famous line 'Philosophers have hitherto only interpreted the world in various ways; *the point is to change it'* [my emphasis]

4 From Engels, F. 'Socialism, Utopian and Scientific', Peking 1975 p86

5 The workers will rightly fight for higher wages though (and better conditions), but simply paying higher wages will not solve the problems of capitalism for workers and their families. As long as capitalism and the profit motive remains, the tendency for the system to go into booms and slumps will remain. The anarchy in the system means that any capitalist not paying the socially necessary wage (ie. paying higher wages than 'socially necessary') will be driven out of business by the competition.

6 It may be stretching a point, but I think it is useful to think of workers also fighting back in both realms – many demonstrations take place in the realm of exchange and strikes, picket lines and occupations (of the workplace) in the realm of production. What brought this to mind was a television documentary of a Brazilian bus worker talking in mid 2013 about how he would be hurrying off to join the massive demonstrations against the government at the end of his shift. I felt the irony of that situation. If the organised bus workers in Rio were to strike and/or commandeer their buses (perhaps to transport demonstrators to the demonstration site), this would be concentrated workers power that would be very difficult for our opponents to better. We

must fight in both realms in a sustained way – the realms overlap – but our power is more concentrated when we act together at work – in the realm of production. And it may be stating the obvious, but we are 1000 times stronger when workers link their struggles with one another.

It wasn't until the textile workers in Egypt, organised, joined the demonstrations in Tahrir square in 2011 that the dictator Mubarak started to topple.

7 The exchange-value of a commodity is proportional to the socially necessary labor time required to manufacture it. Using the socially necessary labor time means using commonly available methodology and tools etc. You cannot give something more value by taking your time to make it.

8 Engels, F. op cit

9 Although the working class must wrest the levers of production from the capitalist class in order to start planning a rational system of production, their seizure of political power will signal the liberation not only of the working class, but of all humanity. For negation of the capitalist class will also negate its twin, the working class. The two classes came into being together and one cannot exist without the other. The wealth and productive forces now liberated will be the possession of all of society. See Engels, F. 'On Marx's Capital', Moscow, p22

10 Tony Cliff was the founding member of the Socialist Workers Party (GB). He and many of the socialist writers and activists who fought alongside him - Chris Harman, Duncan Hallas, Paul Foot referred to here were part of the International Socialist tendency (I.S.). Ian Birchall has written an excellent biography of Cliff. (Birchall, 2011)

11 See Economic Policy Institute report: http://www.epi.org/publication/briefingpapers_bp143/ and See TUC report 'The union Effect' http://www.tuc.org .uk/workplace/tuc-8382-f0.cfm

12 Rebecca Khatun lost her left leg and her right foot at Rana Plaza, the site of the infamous Bangladesh garment factory collapse. That wasn't her hardest trial. Her mother died there. Rebecca has received $120 compensation. On the morning of the collapse, they were worried about the state of the building and didn't want to go in, but were told by the manager 'unless you go in you won't get paid and you'll lose your job' Cockroaches swarm the railing of her bed as another victim explained 'We are poor, we work to live, we entered the factory because we needed to be paid'. See NPR report http://www.npr.org/blogs/parallels/2013/07/10/200644781/Bangladesh-Collapse-The-Garment-Workers-Who-Survived

This reckless approach by capitalists isn't at all restricted to developing countries like Bangladesh

Very recently, a young rail worker in the UK, Scott Dobson, was killed after being hit by a train whilst working long hours for an anti union, blacklisting, casual labor company. 'Accidents' like this are happening all the time. Workers are frightened of being blacklisted if they complain. They are told things like 'you and your gang [ie the workers in the team] – will not get work for a week if you refuse this or that unsafe practice. Breaches of safety rules are a regular occurrence and often lead to injuries and sometimes to deaths at work.

13 Textile and other workers took to the streets.

14 See Cliff, T. 'State Capitalism in Russia'

15 Zero hours contracts causing concern – see: http://www.citizensadvice.org.uk/index/pressoffice/press_index/press09072013.htm

16 Marx, K. op cit, p78

17 Cliff, T. Gluckstein, D. 'The Labour Party: A Marxist History'

Chapter 6
Enter the Dragon

1 see eg., New England Journal of Medicine: http://www.nejm.org/doi/full/10.1056/NEJMra1103676
And National Cancer Institute
http://www.cancer.gov/cancertopics/factsheet/Risk/nuclear-power-accidents
2 Environmental Data Services (ENDS) reports 2001 – 2010

Chapter 7
Some Conclusions

1 A high net worth individual has been defined as anyone with $1 million (£641,000) or more in 'investable assets'. The definition does not include consumer durables such as cars or the value of a main home.
2 The Guardian 19th June 2013
3 I am trying to learn about it though – by studying Marx's Capital etc. We should all try to learn what the capitalist economists are saying so we can see when they are pulling the wool over our eyes.
4 See Nicholas Shaxson, Treasure Islands, Tax Havens and the Men who Stole the World, 2012, Vintage, for an insight into how the capitalists hide the wealth they have stolen from us
5 See John Bellamy Foster's 'Marx's Ecology' for a good explanation of this (New York, 2000)
6 Chris Harman's book 'Zombie Capitalism' (2009) is essential reading if you want to understand the capitalist crisis. See Chapters 12 on for a discussion of the environmental crisis.
7 See Callinicos, A. 'The revolutionary ideas of Karl Marx' (2010)
8 Cliff, T. State Capitalism in Russia (1988 London)
9 Pete Shaw quoted in Jackson (Bookmarks, 2011)

Bibliography

Allen, P., Blake, L., Harper, P. Hooker-Stroud, A., James, P, Kellner, T.,2013, Zero Carbon Britain, Rethinking the Future, CAT Publications

Andrews, D., 2007, 'The potential Contribution of Diesel Standby Generators in Dealing with the Variability of Renewable Energy Sources', in Boyle, G.(ed) Renewable Energy and the Grid, Earthscan, Chapter 7

Birchall, I., 2011, Tony Cliff - A Marxist for His Time, Bookmarks Publications.

Black, M. and Strbac, G., 2005, Value of storage in providing balancing services for electricity generation systems with high wind penetration, Journal of Power Sources

Borer, P., Harris, C, 1998, The Whole House Book, Ecological building design and materials, CAT Publications

Boyle, G (ed), 2007, Renewable Electricity and the Grid, Earthscan

Boyle, G. Renewable Energy, 2004, Power for a sustainable future, Oxford OUP

Brooks, A., Vehicle to Grid Demonstration Project: Grid regulation ancillary service with a battery electric vehicle, 2002 California Air Resources Board

Callinicos, A. 1988, The Revolutionary Ideas of Karl Marx, Bookmarks Publications

Callinicos, A.; Simons, M., 1985, The Great Strike, The miners strike of 1984 and its lessons, Socialist Worker

Cliff, T., 1984, Class Struggle and Women's Liberation, 1640 to the present day, Bookmarks Publications

Cliff, T., 2001, International Struggle and the Marxist Tradition, Selected writings Vols 1 – 4, Bookmarks Publications

Cliff, T., 1988, State Capitalism in Russia, Bookmarks

Crawford, M., 2012, Creating a Forest Garden, Working with

Nature to Grow Edible Crops, Green Books

Dale, L. Milborrow, D., Slark, R., Strbac, G., 2004, Estimates for large scale wind scenarios in UK, Elsevier

Darlington, R., 2013, Radical Unionism, The rise and fall of revolutionary syndicalism, Haymarket, Chicago

Everett, B., 2007, 'Demand Flexibility, Micro-Combined Heat and Power and the 'Informated Grid', in Boyle, G.(ed) Renewable Energy and the Grid, Earthscan, Chapter 8

Ends Report 399, April 2008

Empson, M., 2009, Marxism and Ecology, Capitalism, Socialism and the future of the planet, Bookmarks Publications

Engels, F., 1940, On Marx's Capital, Progress Publishers

Engels, F., 1975, Socialism, Utopian and Scientific, Foreign Languages Press, Peking

Fairlie, S., 2010, Meat A Benign Extravagance, Permanent Publications

Foot, P., 2005, The Vote How it was won and how it was undermined, Viking

Foster, J., 2000, Marx's Ecology, materialism and nature, monthly review press

Gross, R., Heptonstall,P., Anderson, A., Green,T. Leach, J. and Skea, J., 2006, The costs and impacts of intermittency, UK Energy Research Centre, London

German, L., Rees, J., Harman, C., Mcgarr, P.,1994, The Revolutionary Ideas of Frederic Engels, International Socialism Journal

Hallas, D., 1979, Trotsky's Marxism, Bookmarks Publications

Harman, C., 2009, Zombie Capitalism, Global crisis and the relevance of Marx, Bookmarks Publications

Hall, K. (ed), Nichols, R., 2006, The Green Building Bible, The low energy design technical reference (3rd edition) Vol 2

Helweg-Larsen, T. and Bull, J., 2007, ZeroCarbonBritain: An alternative energy strategy, CAT Publications

Helweg-Larsen, T., 2010, The Offshore Valuation, A valuation of the UK's offshore Renewable energy resource, Public Interest Research Centre/Offshore Valuation Group

Infield, D., Watson, S., 2007, 'Planning for Variability in the longer term: The challenge of a truly sustainable energy system', in Boyle, G.(ed) Renewable Energy and the Grid, Earthscan, Chapter 11

Intergovernmental Panel on Climate Change, Climate Change 2007: The Physical Science basis, 2007, Cambridge University Press, Cambridge

Jackson, P., 2011, Close the Gates! The 1972 Miners Strike, Saltley Gate, and the Defeat of the Tories, Bookmarks

Kempton, W. and Letendre, S., 1997, Electric Vehicles as a New Power Source for Electric Utilities, University of Delaware, USA

Kempton, W., Tomic, J., Letendre, S., Brooks, A., Lipman, T., 2001, 'Vehicle-to-Grid-Power: Battery Hybrid and Fuel Cell Vehicles as resources for distributed electric power in California', California Air Resources Board

Kempton, W and Tomic, J., 2004[a], Vehicle to Grid Power Implementation: From stabilising the grid to supporting large scale renewable energy, Journal of Power Sources

Kempton, W and Tomic, J., 2004[b], Vehicle to Grid Power Fundamentals: Calculating capacity and net revenue, Journal of Power Sources, University of Delaware

Kempton, W. and Dhanju, A., 2006, Electric Vehicles with V2G: Storage for large scale windpower, University of Delaware

Laughton, M., 2007, 'Variable Renewables and the Grid: An Overview', in Boyle, G.(ed) Renewable Energy and the Grid, Earthscan, Chapter 1

Lynas, M., 2007, Six Degrees Our future on a hotter planet, Fourth Estate

Martin, C., Goswami, Y., 2005, Solar Energy Pocket Reference, Earthscan

Mathew, S., 2006, Wind Energy: Fundamentals, Resource

Analysis and Economics, Springer
Mackay, D., 2009, Sustainable Energy Without The Hot Air, UIT Cambridge
Mason, P., 2012, Why It's Kicking Off Everywhere, the New Global Crisis, Verso
Marx, K., 1978, Wage Labour and Capital, Foreign Languages Press, Peking
Marx, K., 1975, Wages Price and Profit, Foreign Language Press
Marx, K., Engels, F., 2003, The Communist Manifesto, Bookmarks Publications
Markvart, T. (ed), 2000, Solar Electricity, Wiley
McMullen, R., 2002, Environmental Science in Building, Palgrave
McKillop, A. (ed), 2005, This Final Energy Crisis, Pluto Press
Monbiot, G., 2006, Heat, How to Stop the Planet Burning, Allen Lane
McNeill, J., 2001, Something New Under the Sun, an environmental history of the twentieth century, Penguin
Messenger, R., Ventre, J., 2004, Photovoltaic Systems Engineering, CRC Press
Neale, J., 2008, Stop Global arming change the world, Bookmarks Publications
Patterson, W., 1999, Transforming Electricity, The coming generation of change, Earthscan
Patterson, W., 2007, Keeping the Lights On, Earthscan
Pelsmakers, S., 2012, The Environmental Design Pocketbook, RIBA Publishing
Quaschning, V., 2006, Understanding Renewable Energy Systems, Earthscan
Romm, J., 2005, The Hype about Hydrogen, Fact and fiction in the race to save the climate, Island Press
Scheer, H., 2004, The Solar Economy, Renewable Energy for a Sustainable Future, Earthscan
Scheer, H., 2007, Energy Autonomy, The economic, social and technological case for renewable energy, Earthscan

Shaxson, N., 2012, Treasure Islands, Tax Havens and the Men Who Stole the World, Vintage

Sorensen, B., 2004, Renewable Energy, Elsevier

Stern, N., 2006, The Economics of Climate Change, The Stern Review, Cambridge University Press

Strbac, G., Shakoor, A., Black, M., Pudjianto, D., and Bopp, T., 2006, impact of wind generation on the operation and development of the UK electricity systems

Tomic, J. and Kempton, W., 2007, Using Fleets of electric–drive vehicles for grid support, Journal of Power Sources, University of Delaware

Turton, H. and Moura, F., 2008, Vehicle to Grid systems for sustainable development: An integrated analysis, Technological forecasting and social change

Twiddell, J. and Weir, T., 2006, Renewable Energy Resources, Taylor and Francis

Whitefield, P., 2009, The Earth Care Manual, a Permaculture Handbook for Britain and other Temperate Climates, Permanent Publications

Appendix

1 Rotor length design calculations.

What blade length is required to produce 2.3MW from a wind speed of 12m/s?

Using the equations given by Mackay (2009) p264, B.3 and B.4 gives a blade length of 36m if an efficiency factor of 50% is assumed and 47m if an efficiency factor of 35% is assumed.

The efficiency factor (or power coefficient) of a wind turbine varies depending on a number of design factors – such as blade number, blade profile etc. An efficiency of 50% is a bit optimistic, and most turbines seem to be less efficient than this. Mathews (2006) suggests a figure of 35%. For the purposes of my argument, I have taken an average of the two – giving the blade length of 42m.

2 Power from a given wind speed.

The power (W) obtainable from a given wind speed at hub height is =

Power coefficient x Air density x 3.142 x $(\text{rotor length})^2$ x $(\text{wind speed})^3$ /2

Where the power coefficient is a percentage as discussed above, the density of air is taken as 1.3kg/m^3, the rotor length is in metres and the wind speed in metres/s)

To get a projected hub height speed from the speed at anemometer height, I used an algorithm taken from Mathews (2006) p48

3 Relationship of blade length of a wind turbine to area required – based on the example of the Whitelee windfarm near Glasgow.

We know the land area which has been allocated per turbine, because we know the number of turbines (140) and the land area (55 square kilometres)

So the land area allocated per turbine is 55/140 square

kilometres.

However, for optimum spacing, wind turbines cannot be spaced closer than 5 times their diameter.

That means that each turbine would inhabit an area equivalent to a square with sides of length equal to 5 times the turbine diameter – an area of $(5d)^2$ where d is the turbine diameter.

We can now get the optimum blade length for a situation where we are placing 140 turbines in an area of 55 square kilometres.

The turbine blade length is half the turbine diameter (i.e. the radius of the circle the blade sweeps through)

So, if r is the radius in metres, then

$(5 \times r \times 2)^2$ = allocated area per wind turbine = $55/140 \times 10^6$

r = 100 x m

Blade length, r = 62.67m

4 The area taken up by a windfarm, given the blade length (r) and the number of optimally spaced turbines (n) is given by:

Area = $r^2 \times n \times 100$

(nb If the area is in km^2, the radius will be in km)

5 Some energy units and conversion factors:

Powers of ten:

Prefix, Symbol, Value:

kilo, k,10^3

mega, M.10^6

giga, G,10^9

tera,T, 10^{12}

Units:

Power: watts, kilowatts, megawatts etc; Symbols W, kW, MW etc respectively

Energy: watt hour, kilowatt hour, megawatt hour; Symbols Wh, kWh, MWh etc. respectively

The UK Digest of Energy Statistics (DUKES) uses the energy unit : Tonnes of oil equivalent , (toe)

In the US, frequently, British Thermal Units (Btus) are used

1kWh = 8.6×10^{-5} toe = 3412 Btu

1 toe = 12000 kWh = 3.9×10^{7} Btu

1 Btu = 2.52 toe = 2.9×10^{-4} kWh

Tables

Daily Energy Balance

Hour	Total demand in MW - Electrical, transport and heating/industrial (reduced from current levels by 45% efficiency)	Renewable power produced (Mainly offshore wind) MW	Vehicle to Grid fleet charge %	Energy harnessed but not consumed or stored. MWh
0	85808	89060	99.2	0
1	85217	87206	99.3	0
2	84855	98098	100	4247
3	84462	94251	100	9789
4	57209	84426	100	27218
5	57610	87069	100	29459
6	59376	69813	100	10436
7	62811	62193	100	0
8	64952	69954	100	4385
9	65782	64091	99.9	0
10	65976	78748	100	11081
11	66212	51375	98.9	0
12	66381	64675	98.8	0
13	66175	54096	97.9	0
14	65836	93869	100	0
15	65573	87107	100	20946
16	66058	73897	100	7839
17	66302	54138	99.1	0
18	65513	42177	97.4	0
19	65073	36123	95.3	0
20	66545	43954	93.6	0
21	65769	27904	90.8	0
22	90727	35522	86.8	0
23	87880	29797	82.5	0

Table 1

The above table (Table 1) shows an example daily energy balance with a system comprising 44250 10MW offshore wind turbines (plus the Severn Barrage) supported by 27.2 million electric vehicles providing 1360GWh of electrical storage.

It can be seen that, when the fleet is fully charged, there is a surplus of energy that cannot be stored.

The electric transport is charged late at night and in the early hours of the morning. This explains the greater electrical demand at those times.

The numbers arrived at are precise because I have used exact figures for the energy demand in 2007 for the UK (from the Digest of UK Energy Statistics - DUKES) and I have used some precise numbers for the wind speeds. In practice, of course, we would not know the precise energy demand figures in advance, nor would we know the precise wind speeds. But because the variability patterns of the wind will be more or less the same from one year to the next, and because the energy demand figure will not change radically from year to year, this approach will provide a good rough approximation for future infrastructure requirements.

Timeline

Year	Number of 1(MW Wind-turbines Installed	Millions of / GWh of Electrical Storage	Wind Energy Produced TWh	Millions of of CO2 Emissions
0	0	0	0	447
1	500	1.6/80	21	427
2	1000	3.2/160	43	407
3	2000	4.8/240	83	382
4	3500	5.4/270	149	352
5	5000	7.0/350	206	322
6	7000	8.6/430	268	289
7	9000	10.2/510	345	253
8	11400	11.8/590	465	204
9	14000	13.0/650	579	162
10	17000	14.6/810	720	124
11	20000	16.2/810	847	101
12	23000	17.8/890	978	80
13	26000	19.4/970	1102	66
14	30000	21.0/1050	1265	55
15	34000	22.6/1130	1452	45
16	37000	24/1200	1578	38
17	41000	25.6/1280	1730	31
18	44250	27.2/1360	1873	0

Table 2

The above table (Table 2) shows, year by year, how the CO_2 emissions are gradually reduced as more and more offshore wind turbines are installed. Finally after around 19 years of sustained

manufacture and installation of generation, storage and transport infrastructure, they are eliminated.

Timeline Continued

Table A.3

Year	Fossil fuel bur TWh	Wind energy TWh Not stored or consumed	Tidal-Range pow TW (Severn Barrage
0	1594	0	0
1	1523	0	0
2	1451	0	0
3	1361	0	0
4	1260	0	0
5	1153	0	0
6	1040	0	0
7	913	0	0
8	743	1	0
9	601	17	0
10	468	75	0
11	386	158	13
12	309	286	13
13	259	411	13
14	216	581	13
15	180	769	13
16	152	926	13
17	122	1100	13
18	0	1263	13

Table 3

This table (Table 3) is an extension of Table 2. It shows year by year, the reduction in fossil fuel burn, and it also shows the contribution of the tidal range electricity generator which could be built on the Severn estuary near Bristol in the UK. The tidal range barrage contributes a significant and predictable amount of energy, but it is several years before it comes on line due to the enormous building challenge the Severn Tidal Barrage presents.

Earth Books are practical, scientific and philosophical publications about our relationship with the environment. Earth Books explore sustainable ways of living; including green parenting, gardening, cooking and natural building. They also look at ecology, conservation and aspects of environmental science, including green energy. An understanding of the interdependence of all living things is central to Earth Books, and therefore consideration of our relationship with other animals is important. Animal welfare is explored. The purpose of Earth Books is to deepen our understanding of the environment and our role within it. The books featured under this imprint will both present thought-provoking questions and offer practical solutions.